SCIENCE UNDER SCRUTINY

AUSTRALASIAN STUDIES IN HISTORY AND PHILOSOPHY OF SCIENCE

VOLUME 3

SCIENCE UNDER SCRUTINY

The Place of History and Philosophy of Science

Edited by

R. W. HOME

Department of History and Philosophy of Science
University of Melbourne, Australia

D. REIDEL PUBLISHING COMPANY

A MEMBER OF THE KLUWER ACADEMIC PUBLISHERS GROUP

DORDRECHT / BOSTON / LANCASTER

Library of Congress Cataloging in Publication Data
Main entry under title:

Science under scrutiny.

    (Australasian studies in history and philosophy of science; v. 3)
    Includes bibliographical references and index.
    Contents: Introduction / Lloyd Evans – Why philosophy of
science? / John Passmore – Knowledge and power in the sciences /
Everett Mendelsohn – [etc.]
    1. Science–Philosophy–Addresses, essays, lectures. 2. Science–
History–Addresses, essays, lectures.   I. Home, Roderick Weir.
II. Series.
Q175.3.S323   1983   501   83–17684
ISBN 90-277-1602-1

Published by D. Reidel Publishing Company,
P.O. Box 17, 3300 AA Dordrecht, Holland.

Sold and distributed in the U.S.A. and Canada
by Kluwer Academic Publishers,
190 Old Derby Street, Hingham, MA 02043, U.S.A.

In all other countries, sold and distributed
by Kluwer Academic Publishers Group,
P.O. Box 322, 3300 AH Dordrecht, Holland.

Printed in The Netherlands.

TABLE OF CONTENTS

FOREWORD

Only in fairly recent years has History and Philosophy of Science been recognized — though not always under that name — as a distinct field of scholarly endeavour. Previously, in the Australasian region as elsewhere, those few individuals working within this broad area of inquiry found their base, both intellectually and socially, where they could. In fact, the institutionalization of History and Philosophy of Science began comparatively early in Australia. An initial lecturing appointment was made at the University of Melbourne immediately after the Second World War, in 1946, and other appointments followed as the subject underwent an expansion during the 1950s and '60s similar to that which took place in other parts of the world. Today there are major Departments at the University of Melbourne, the University of New South Wales and the University of Wollongong, and smaller groups active in many other parts of Australia, and in New Zealand.

"Australasian Studies in History and Philosophy of Science" aims to provide a distinctive publication outlet for Australian and New Zealand scholars working in the general area of history, philosophy and social studies of science. Each volume will comprise a group of essays on a connected theme, edited by an Australian or a New Zealander with special expertise in that particular area. The series should, however, prove of more than merely local interest. Papers will address general issues; parochial topics will be avoided. Furthermore, though in each volume a majority of the contributors will be from Australia or New Zealand, contributions from elsewhere are by no means ruled out. Quite the reverse, in fact — they will be actively encouraged wherever appropriate to the balance of the volume in question.

R. W. HOME
General Editor
*Australasian Studies in History
and Philosophy of Science*

PREFACE

'History and Philosophy of Science' is the name most commonly used these days to refer to that form of scholarly activity which takes science itself as the object of its inquiry. The subject embraces a wide diversity of intellectual styles. Some of its practitioners bring to their work a particular interest in historical questions, others devote themselves primarily to philosophical issues, others again approach the subject from a generally sociological point of view. While the concerns of some are scholarly and abstract, others deal more with the practicalities of the here-and-now, for example by advising governments on matters of science policy or by fostering a heightened public awareness of the environmental and ethical dilemmas posed by modern science and its applications. Some come to the field with a strong background in one or other of the sciences, others approach the subject as outsiders to science who, recognizing how powerful an influence this is on contemporary society, seek a better understanding of its internal dynamics and modes of operation.

These diversities of background and interest cannot, however, obscure the fact that the work of those involved also has many interconnections, sufficient indeed to ensure that, despite occasional differences of opinion concerning nomenclature, we are really dealing here with a single field of intellectual endeavour rather than with several.[1]

Furthermore, it is a field that has undergone dramatic growth during the past two or three decades, in the course of which both its style and its intellectual content have been transformed. Now, however, in a period of increasing budgetary constraint in most parts of the world, that growth has largely ceased; indeed, in a number of countries the discipline today finds itself uncomfortably exposed to possible cut-backs rather than continued expansion. There has been widespread agreement among those engaged in the field as to the potential utility of their work, whether in helping to clarify some of the outstanding issues confronting modern society or in contributing to the education of its citizens. However, many of the new perspectives that have been achieved have yet to receive general recognition in the community at large, and considerable uncertainty exists as to how best to bring the work that is being done to the attention of the thinking public. This is therefore

R. W. Home (ed.), Science Under Scrutiny, ix–xvii.
© 1983 by D. Reidel Publishing Company.

a particularly opportune time to undertake a reassessment of such studies, more especially with regard to their educational role and bearing on contemporary problems.

These problems exist in Australasia as elsewhere, and for this reason the Australian Academy of Science readily agreed to a proposal from its National Committee for History and Philosophy of Science that it sponsor an international symposium in which the issues involved could be explored. This was held at the University of Melbourne in August 1979. The principal papers presented are brought together here in published form in order to make them available to a wider audience.

During the first half of the twentieth century, science was seen in Western societies in almost wholly optimistic terms. That orderly system of inquiry into the operations of nature which Francis Bacon had envisaged as the key to the 'improvement of man's estate' seemed at last to have been established. The successes of science and science-based technology in ameliorating the conditions of life led to a widespread faith in 'the scientific method' as a means of problem-solving that could be applied well beyond the bounds of science itself. An early function envisaged for university courses in History and Philosophy of Science was the imparting of this method, the correct characterization of which was not regarded as in any way problematic, to non-science or beginning science students. Beyond this, the chief role of History and Philosophy of Science was to be simply to celebrate the achievements of the great scientists of the past. Numberless histories were written in which the major scientific discoveries were carefully catalogued and ascribed to those regarded as responsible for them. The main concerns of the authors of such works were completeness of coverage and correctness of attribution. The search for precursors was a leading preoccupation of the discipline as they understood it.

Inevitably, the better scholars engaged in the field did not remain satisfied with such an approach. Leading philosophers of science presented analyses that recognized and attempted to take adequate account of the complexities of scientific argumentation; leading historians of science took seriously the intellectual concerns of the scientists of the past whose writings they were studying, and attempted to construct genuinely historical accounts of their work. Yet the discipline remained fundamentally uncritical in the sense that science itself remained beyond criticism, a standard against which other forms of inquiry were to be measured. The principal objective of the logical positivists, for example, who dominated Western philosophy of science for a generation, was to promote a properly empirical 'scientific outlook' against the rival philosophical claims of idealism and transcendental metaphysics. Likewise, a generation of historians of science shared a conviction that science was the most remarkable achievement of Western intellectual history, and agreed with Charles C. Gillispie in depicting its growth in terms of an advancing 'edge of objectivity'.[2]

More recently, science has been displaced from this lofty pedestal, and in the eyes of many its image has become tarnished. In an age of nuclear weapons, omnipresent insecticides and other industrially produced poisons, and threats of mass unemployment brought about by science-based technology, many have come to fear the products of science as much as they once looked forward to them. In many quarters, science is seen no longer as an innocent and dedicated search after truth, but as an integral and therefore tainted part of a discredited economic system. The lonely investigator driven by an insatiable curiosity to know has been reduced to a mere cog in a world dominated by 'industrial science'.

At the same time, the concerns of the scientist remain in a very real sense isolated from the rest of modern intellectual life. The profound division within educational programmes everywhere and at every level between science and the other components of the curriculum both reflects and ensures the persistence of a common perception of science as a thing alien to and remote from man's other intellectual activities. Science is seen as mysterious, as 'too hard' for ordinary mortals to cope with. But now the friendly image of the absent-minded professor who can cope with such matters has been replaced in the popular imagination by the cold and distant figure in a starched white laboratory coat who manipulates his arcane yet powerful knowledge in a distinctly threatening way. Such images, threatening or not, are of course mere caricatures, but they do point up the yawning gap in understanding between the sciences and the humanities, C. P. Snow's 'two cultures', to which Lloyd Evans refers in his introductory remarks below.

One function that has often been claimed for studies in History and Philosophy of Science is that they can help to bridge this gap. The problem has been to design suitable courses that would achieve this: not even the most successful of those that have been tried so far have fully lived up to the hopes of those promoting them. At the professional level, historians and philosophers of science for at least a generation now have been more concerned to strengthen the foundations of their own discipline than to use their peculiarly interdisciplinary position to promote discourse between their scientist and humanist colleagues. Yet as the power of science to shape society grows inexorably, the need to bridge the gap between our leading groups of intellectuals becomes ever more pressing, and History and Philosophy of Science remains one of the most likely aids to doing so.

In the process of becoming industrialized, science has come not only to have more impact on society but also to cost considerably more than it once did. For both reasons, and especially because much of the money involved inevitably comes either directly or indirectly from public sources, governments everywhere have found it increasingly necessary to develop policies for the organization and funding of research and development. Here, too, History and Philosophy of Science has begun to find a role, since, in principle at least, people working in this discipline ought to have a good understanding of the nature and workings of science, and yet at the same time their advice ought to be free, as that of practising scientists might not be, of any suspicion that it represented merely the vested interest of their own profession. As both Lloyd Evans and Jarlath Ronayne indicate in their contributions to this volume, the sub-discipline of Science Policy Studies has thus far led a stormy existence. Ronayne argues, however, that the field has now established a widely agreed framework for discussion. He points to the strong orientation to policy studies in several Australian university departments of History and Philosophy of Science as part of a 'significant Science Policy Studies infrastructure' that has been built up in this country in recent years, and expresses the hope that, here as elsewhere, governments are beginning to look to sources of this kind for informed advice.

The transformation that has occurred in recent years in the style and content of work being done in History and Philosophy of Science has parallelled and perhaps even contributed to the change in attitude towards science itself. In philosophy of science, it is the very objectivity traditionally claimed for scientific discourse that has come under challenge. In particular, the once sharp distinction between observation and theory in science has come to be widely questioned. Observation reports, we are now assured by many, are irredeemably theory-laden: no longer seen as objectively true statements about the world, they therefore cannot be used, or so it would appear, as evidence on the basis of which we might distinguish between different very general scientific theories. If this be true, it suggests, as John Passmore notes, that "one of the things that philosophers of science have traditionally tried to do, namely to show that science rests on firm foundations, is in principle impossible". It throws into doubt many of the claims that have traditionally been made for science as embodying a privileged kind of knowledge. It leads us to ask, with Paul Feyerabend, the question addressed by Alan Musgrave in his paper in this volume, 'What's so great about science?'.

Musgrave does not, in fact, try to establish an answer to this question, though he does suggest one, namely "what is great about science is its ability to make intellectual progress". Instead, he confronts what he conceives as a crucial failing in much contemporary philosophy of science, its neglect of the dichotomy between facts and values. This dichotomy is, Musgrave asserts, one of the "very few solid discoveries" that philosophy has to its credit. It is not the same thing as the once-popular dichotomy between fact and theory, and is in no way overturned if the latter is abandoned. In neglecting it, philosophers of science have created needless difficulties for themselves. In particular, it is essential to take it into account in any attempt to characterize 'intellectual progress'. By himself applying the principle rigorously, Musgrave constructs a powerful critique of much of the recent literature on this subject.

Yet Hugh Stretton's paper in the present collection shows that the fact/value dichotomy is not itself immune from criticism, at least as it has been applied (or misapplied) in the social sciences. In typically lucid style, Stretton delivers a devastating attack on much of what has been written in these disciplines — especially sociology and economics — in the past generation. The majority of social scientists, he argues, have mistakenly understood the distinction between facts and values as an injunction to refrain from injecting value judgements into their work. Accordingly, they have sought to confine themselves to describing society as it is or might be as a result of different policy choices, and have professed to leave the actual choices of policies to others. The result, says Stretton, was a science as sterile as it was reactionary. Sociology became more and more inward-looking and less and less concerned with the society it claimed to describe, economics degenerated into abstract modelling that paid heed solely to economic rather than social or political causes of economic effects, and thus became systematically blind to economic subjects of social concern. "Social science," Stretton maintains, "is above all a *recombinant* science". The social scientist is inextricably enmeshed in the social system of which he is a member, and as a result his values are inextricably enmeshed in his science. 'Value-free social science' is a myth, a science in which the values of the scientists have merely been concealed. Stretton advocates instead, with Martin Rein, a form of inquiry in which "not only are values treated as the subject of analysis, but it is assumed that analysis can never be independent of the values we hold". That is, Stretton wishes to see discussions of values reintroduced into the classroom where they belong instead of being treated as if they form no part of the science in question.

Rom Harré in his paper reaches somewhat similar conclusions with regard to psychology. He argues that this science, too, in its traditional form, was sterile, and for the same reasons that Stretton gives in the cases of sociology and economics. He maintains that the supposed 'objectivity' of traditional psychology amounted merely to a failure to recognize that the subject rested on various unstated assumptions about human behaviour that were in fact highly culture-specific and value-laden. Worse, these presuppositions carried with them a number of objectionable moral and political implications. In the 'new psychology' which Harré advocates, human beings are no longer treated as automata but as active agents whose interpretations of the events in which they participate — whose values, in other words — necessarily form a central part of the psychologist's concern. "Those," writes Harré, "who used heretofore to be called, contemptuously, 'subjects', enter the scientific research teams of the new psychology as members in good standing. And this involves something morally and politically potent, namely listening to and taking seriously what the folk have to say about their own actions."

Among historians of science, too, the objectivity of science has in recent times been strongly challenged. A generation ago, the literature of the history of science was devoted almost exclusively to affairs of the intellect, questions 'internal' to scientific thought rather than 'external' factors whether social, economic or political. Furthermore it was generally maintained, at least in the Western literature of the subject, that the history of scientific ideas was a relatively autonomous discipline that could be studied in relative isolation from all external, 'non-scientific' considerations. Now, however, the old argument of Marxists and others that the content of science is inextricably linked with its social context finds a more receptive audience, even though in practice it has proved extremely difficult to specify the nature of the links involved. While most continue to reject the claim of the sociologists of knowledge actually to explain the invention of particular scientific ideas in terms of social factors of one kind or another, there has been a general swing towards portraying the evolution of science in terms of its mutual interactions with the society in which it is embedded rather than from a purely intellectual point of view. This goes well beyond recognizing that social factors can determine which kinds of questions are taken up by scientists at any particular time — this has, in fact, never been disputed — to the assertion that, very often, the most interesting and important questions that need to be answered by historians of science concern social rather than purely intellectual matters. In particular, as science has increasingly come to be a powerful force for good or evil in the world, it has come to be felt more and more that an essential part of the historian's task — perhaps *the* most important part of his task — is to show how this has happened. And this is, of course, largely a social question, even if intellectual developments remain an important part of the story.

Everett Mendelsohn in his paper in this volume addresses precisely this question, and in doing so illustrates very clearly the new approach to the subject as a whole. Mendelsohn's concern throughout is with science as a social institution within the wider context of Western society in general. He traces both changing social structures within science itself – noting, for example, the growing dominance of 'pure' over 'applied' science during the course of the nineteenth century, precisely at the time when science first began to play a significant industrial role – and the way in which science as a growing locus of power came to be accommodated within the wider social system. He shows how the present separation between scientists and general public developed as science became professionalized, and concludes that in this new age of spreading disillusionment with science, History and Philosophy of Science can play an important part in the work of reconstruction that is necessary.

This is but one of the roles for History and Philosophy of Science suggested by contributors to the present volume. John Passmore notes several others in his masterly survey of the different categories of philosophy of science. What he calls 'substantive' philosophy of science, for example, contributes directly to clarifying substantive problems within science itself. On the other hand, what he calls 'critical' philosophy of science serves to confine the claims of the sciences to their proper boundaries. Just so, Harré details the use of History and Philosophy of Science as 'science criticism' in his attack on the claims of the behaviourist psychologists, and shows how it was with its aid that the 'new psychology' he describes was born. Just so, too, Stretton exposes the limitations of traditional sociology and economics. Finally and most importantly, Passmore's 'social' philosophy of science raises and attempts to clarify a set of moral problems that neither scientists nor laymen can afford to ignore.

Bryan Gandevia pursues a rather different theme. He shows by means of some telling early Australian examples that the history of science, and more particularly the history of medicine, has an important contribution to make to our understanding of social history. Epidemiological studies of the health and disease characteristics of populations are, he argues, fundamental to any worthwhile social history, because they are the primary indicators of how men adapt to their environment in a particular place at a particular time. The specialist medical historian is, he points out, especially well fitted to undertake investigations of this kind.

Gandevia also discusses, rather more briefly, the place of medical history within a general medical education. In this, his paper reflects a pervasive theme of the volume as a whole, namely, the educational value of studies in History and Philosophy of Science. Differences of opinion emerge, however, as to how the discipline might best fulfil its role in this regard. Harré wants History and Philosophy of Science to be taught as a resource, a basis for an on-going practice of 'science criticism'. He doubts, however, whether simply teaching the subject as an independent part of the school curriculum will achieve the result he desires. "HPS taught as a topic," he observes, "is not easily transformed into a resource." Harré's preferred alternative is to teach it "where teachers are taught, that is, in universities and above all in teacher training colleges". Passmore agrees: "No science educator," he writes, "can properly ignore the issues raised by philosophy of science. In any system of teacher training, 'philosophies-of' ought to play a central role".

Passmore also believes that "no science student should leave school and university without having some sense of science as a living, active process of inquiry, not just as a set of formulae and professional tricks". In this he is joined by Musgrave, who argues that the all too typical 'textbook indoctrination' to which science students are exposed "conveys neither a proper understanding of current science nor a proper appreciation of what is great about it". If what is great about science is, as Musgrave supposes, its ability to make genuine intellectual progress, and if we want students studying science to come to understand this, then they must be given a real sense of the progressive nature of the subject. But, as Musgrave points out, for this, "past science must figure as well as current orthodoxy"; in other words, the science curriculum must include a genuinely historical component.

Randall Albury arrives at a much more radical conclusion which many a science educator will find quite alarming. He proposes a systematic confrontation in the classroom between common sense and scientific knowledge, not in order to show the superiority of one approach above the other but "in order to determine the forms of production and the legitimate domains of application of each". Albury insists, however, that if this is to be achieved, the concepts of both common sense and science must be presented from a radically instrumentalist point of view, stripped of any and all commitments concerning the real natures of things. In this way, he claims, we would arrive at an integrated curriculum, combining social studies and natural science: "neither scientific knowledge nor common sense knowledge would be devalued, but each would be assigned its appropriate place in the student's intellectual repertoire".

Many difficulties, philosophical, social and practical, stand in the way of such a scheme. By its very radicality, however, it may startle science educators into reflecting more deeply upon both their objectives in their teaching and the interests which this represents. As Stretton so forcefully reminds us, we cannot avoid making value judgements, even in relation to our science. Intelligent discussion of these is surely what is needed in our science curricula, as well as in our laboratories and government agencies.

R. W. HOME
University of Melbourne

NOTES

[1] Alternative names that have been used to embrace some or all of the interests I am here subsuming under History and Philosophy of Science include Science Studies, Liberal Studies in Science, Social Studies of Science, History and Sociology of Science, Science Policy Studies, Science of Science, and the technically precise yet ugly Meta-scientific Studies.

[2] Cf. Gillispie, (1960) *The Edge of Objectivity: An Essay in the History of Scientific Ideas* (Princeton).

LLOYD EVANS

INTRODUCTION

When the natural sciences began to bloom in the seventeenth century, the scientists rather forcefully demarcated themselves from the more traditional fields of learning. The choice of motto by the Royal Society has been viewed as a deliberate rejection of humanist learning, as has Robert Hooke's draft preamble to its statutes which reads: "The business and design of the Royal Society is − To improve the knowledge of natural things . . . (not meddling with Divinity, Metaphysics, Moralls, Politicks, Grammer, Rhetorick, or Logick)". History, you will have noted, was not mentioned in Hooke's list.

There were a number of reasons why the early scientists separated themselves from the humanities.[1] The latter, naturally enough, responded in kind, and our respective branches of learning largely went their separate ways, in Australasia no less than elsewhere. Our earliest scientific society, founded in 1821 as The Philosophical Society of Australasia, fined its members heavily if they mentioned polemical divinity or party politics.[2] On the other hand, I notice that the current constitution of the Australasian Association for the History and Philosophy of Science does not include practising scientists in the list of categories it recognises as being professionally involved in such studies.

George Seddon has pointed out that history courses in Australian universities, even one entitled 'Seventeenth century revolutions in Britain', usually make no mention of science,[3] although science has been said to be "the most dynamic, distinctive and influential creation of the Western mind".[4] Historians and their students have lost by this estrangement, and so have the scientists, who could profit greatly from the insights and perspectives of historians of science. To give a personal example, I have found Oliver MacDonagh's paper on 'Government, Industry and Science in Nineteenth Century Britain'[5] highly enlightening on the subject of how the class distinction between pure and applied science developed in Britain. This may be history, but the attitudes it deals with still influence research policy, and we gain by understanding their origins. We also gain by having so judicious and understanding a philosopher as John Passmore to give perspective to the criticisms and concerns to which science and scientists are subject.[6]

As someone who was lucky enough to be lectured to as an undergraduate

1

R. W. Home (ed.), Science Under Scrutiny, 1−3.
© *1983 by D. Reidel Publishing Company.*

by Karl Popper, in his New Zealand days, I believe we scientists lose a great deal from being generally — excepting the Bernals, Polanyis and Medawars amongst us — distant from the philosophers of science. Equally, had you been closer to practising scientists, perhaps you would have placed less emphasis on the heroic personalities, innovations and paradigms of 'aristoscience' as Passmore has called it, and rather more on Kuhn's 'normal science'.[7] I might add that many biologists and earth scientists consider physics to have played too prominent a part in the philosophy of science, but perhaps molecular biology and plate tectonics[8] are already changing that. Molecular biology, for example, is generating not only new combinations of DNA, but also new ethical problems, as discussed in a recent issue of *Daedalus*.[9] Perhaps this is why genetic engineering has twitched the public nerve, because it is here that science begins to touch upon the really fundamental concerns of mankind.

The gist of what I have been saying, I suppose, is that one of the major educational roles of HPS studies is probably in the mutual education of historians and philosophers on the one hand and of practising scientists on the other. The process has always gone on, but such is the impact of science now that it needs amplification. It is important that the young be exposed to this interactive process, as also students from less developed countries, if both science and the study of it are to be healthy components of our life and culture.

This brings me to their bearing on contemporary problems, the second theme of this symposium and the particular concern of the third component of the International Union of the History and Philosophy of Science, namely Science Policy Studies, and its International Council (ICSPS). Neither the scientific nor the humanitarian parent has known quite how to handle this vigorous and occasionally frustrated adolescent. At times he has seemed to suffer from an Oedipus complex, disposing of one parent and amalgamating with the other, but some of the home truths about science for which we are indebted to science policy studies have been helpful. Practising scientists do not relish the kinds of analysis, indexation, scrutiny and control to which their research is increasingly subject, and I believe we must fight to retain certain autonomies if the creative spirit of science is not to be crushed. Science is, and should be, fun. In the words of Thomas Sprat,[10] "The *Works of Nature* . . . are one of the best and most fruitful soils for the growth of Wit", and it would be a pity if science lost that zest through excessive control. Equally, however, we have to recognize the enormous, but often dimly foreseen, impact of science on daily life, which requires science to renegotiate its contract with society, and with other forms of learning. Science policy

studies will surely help in what may be, for some time yet, a painful process
of re-thinking our scientific ethic.

Australian Academy of Science

REFERENCES

[1] Ben-David, J. (1977) 'Organization, Social Control and Cognitive Change in
     Science', in *Culture and Its Creators*, eds. J. Ben-David and T. N. Clark, Chicago:
     University of Chicago Press, pp. 244–265.
[2] Moyal, A. (1976) *Scientists in Nineteenth Century Australia: A Documentary
     History*, Melbourne: Cassell.
[3] Seddon, G. (1973) 'The Teaching of History in Australian Universities', unpub-
     lished ms.
[4] Butterfield, H. (1949) *The Origins of Modern Science 1300–1800*, London: Bell
     and Sons.
[5] MacDonagh, O. (1975) 'Government, Industry and Science in Nineteenth Century
     Britain: A Particular Study', *Historical Studies* 16, 503–517.
[6] Passmore, J. (1978) *Science and Its Critics*, London: Duckworth.
[7] Kuhn, T. S. (1970) *The Structure of Scientific Revolutions*, 2nd ed., Chicago:
     University of Chicago Press.
[8] Frankel, H. (1979) 'The Career of Continental Drift Theory', *Studies in History
     and Philosophy of Science* 10, 21–66.
[9] *Daedalus*, Spring 1978.
[10] Sprat, T. (1667) *The History of the Royal Society of London*, London: J. Martyn.

JOHN PASSMORE

WHY PHILOSOPHY OF SCIENCE?

Most of the time, we engage in activities without asking ourselves why we are doing what we are doing. Whether as philosophers, scientists, politicians, teachers, doctors, mechanics or artists, we inherit a complex tradition. We seek to solve problems which satisfy that tradition's concept of what it is to be a problem; we tackle such problems by professionally approved methods; our aim is to find a solution which will stand up to the sort of criticism our problem-solving community takes seriously. Of course, such traditions are neither monolithic nor unchanging. In and of themselves, they generate dissidents, heretics, whose unorthodoxies for the most part fade into oblivion as 'lost causes' but sometimes turn the tradition in a new direction. The traditional concept of what constitutes a problem may then alter, accepted procedures may be modified, new forms of solution may be demanded. The locus of seriously-taken criticism may correspondingly shift. But for the most part, the participants in an activity can accommodate themselves to quite revolutionary changes without being forced to reconsider the general thrust of their activity or, even, its professional standards. Although Einstein's relativity theory, to say nothing of quantum mechanics, came as a shock, it in no way disrupted the tradition of physical science. Or, at least, neither innovation so disrupted that tradition as to compel physicists to ask themselves: 'Why are we doing physics?'

Compared with science, philosophy is a relatively loose tradition. Philosophers, considered internationally, disagree far more than do scientists about the character of their problems, the methods of procedure proper to philosophy, what counts as a solution, what kinds of criticism have to be taken seriously. In consequence they are much more inclined than are scientists to ask themselves why they are doing what they are doing, what philosophy is 'about'.

This, it might be argued, is a clear sign that science is in far better health than is philosophy. Philosophers, on this view, are intellectual hypochondriacs, just because philosophy, for all its long history, has never established itself as a tightly-regulated discipline. Every now and then a philosopher announces that he has at last succeeded in converting philosophy into a science-like inquiry. But somehow the science-like discipline — be it phenomenology

5

R. W. Home (ed.), Science Under Scrutiny, 5–29.
© *1983 by D. Reidel Publishing Company.*

or dialectical materialism — remains culture-bound or, at best, confined to a relatively small group of international enthusiasts.

Then how can so insecure a discipline as philosophy expect that scientists should pay any attention to it? It is easy enough to understand why W. H. Watson writes as he does:

> The soul of physics is given by physicists who think about it, who do experiments, discuss it, write about it and teach it. This is the only kind of soul worth having. The rest is a sort of pathological morbidity that keeps a man from learning about nature, and discourages real participation in that creative process. In good health it is not natural. Philosophy, as Wittgenstein once remarked, ought to liberate us from the idea that there is a kind of academic doctor who can do things for physicists and other scientists that they are incapable of doing for themselves.[1]

To any such a philosophical doctor, the scientist will be more than tempted to reply: 'Physician, heal thyself'. The only proper function for the philosopher of science, one might well conclude, is to bow himself out, to demonstrate that science is for scientists, and for them alone, and then to lapse into silence. 'What right have you', I have been asked in the same spirit by a distinguished historian, referring to my work in the philosophy of history, 'to tell historians what they should do?' And this although I was writing, I thought, a defence of history, by no means attempting to legislate. Philosophy of history is, in his view, a form of meddling. Many scientists, I suspect, feel exactly the same way about the philosophy of science.

We can certainly agree with Watson that it is not healthy to be perpetually asking ourselves, whether in philosophy or outside it, why we are doing what we are doing: that way lies catatonia. 'Get on with the job' is, in general, a fruitful imperative. But not always. For one can be locked by a tradition into professional attitudes, professional prejudices, professional judgments of competence, in a manner which is both destructive to the imagination — and hence to the progress of the activity in which one is engaged — and insensitive to the needs of those outside the profession. Many of us would say that something of this sort is now true of the medical profession, under the influence of so-called 'scientific medicine'. And some, if a smaller number, would say that it is true of science. Karl Popper has forcibly condemned what he calls 'an obscurantist faith in the expert's special skill and his personal knowledge and authority'. That obscurantist faith is particularly likely to arise within professions which disdain all criticism except 'peer-criticism'.

The looseness of philosophy, one should add, is not necessarily a sign of ill-health. Paradoxically enough, this very looseness has enabled philosophy to give birth to stricter disciplines — or at least powerfully to contribute to their birth — and to generate fresh attitudes to the world, including human society. It has permitted, too, the penetration into philosophy of those who were not at first trained in that field, Frege the mathematician or Wittgenstein the engineer or Popper the school-teacher. It does not follow, as Paul Feyerabend has suggested, that science should imitate philosophy's looseness, taking as its model the pre-Socratics, with their competing world-views, rather than post-Newtonian professional science. Its relative rigidity may be as important to the growth of science as is a relative looseness to the growth of philosophy. But not if the rigidity is of a sort which simply rules out of court any questions whatsoever about science's problems, its methods, its accepted doctrines, or its character as a form of life. That way lies fossilisation. It is as unhealthy for an activity to try to protect itself against the question 'Why?' as it is unhealthy for it to be always asking about itself 'Why?'

Sometimes, indeed, an activity has no option. The question 'Why?' may be forced upon it by harsh critics. So, earlier in this century, philosophers were told by the logical positivists, most of them scientists by training, that many of the problems with which they had traditionally concerned themselves were pseudo-problems, that they were tackling such of their problems as were not pseudo-problems in quite the wrong way, and that what they counted as 'good solutions' were in fact meaningless. Faced with such objections, they were compelled to ask themselves: 'Why philosophy?' Philosophers of science, more recently still, have been subjected to equally basic criticisms. They have been told that one of the things they have classically tried to do, namely to show that science rests on firm foundations, is in principle impossible; they have been told that they have been proceeding in quite the wrong way, looking to logic when they should have been looking to history; they have been told that their 'philosophy of science' is in fact quite inapplicable to actual science, that such value as it has is merely as a technical exercise. Under these circumstances, they, too, cannot but ask themselves: 'Why philosophy of science?'

There are other circumstances, besides such fundamental attacks upon its very right to exist, in which self-scrutiny can be forced upon an activity. Its proponents may claim that it has been unduly neglected, that it ought to be taught in schools, ought to be given wider recognition, ought to be awarded research grants. So philosophers of science now sometimes urge that their subject should be taught to science students, whether at school or at universities, that it has at least as strong a claim as history of science to be recognised by such bodies as the Royal Society, that it should be supported by Science Research Foundations. Science itself once had to justify its claims to a place in our

education system, and now has to justify its ever-growing demands for
financial support. That has forced upon it a degree of self-consciousness.
The temptation under such circumstances is to exaggerate by claiming for
an activity publicly-attractive virtues which it does not in fact possess, as
when scientists exaggerate the likelihood that their inquiries will have important technological spin-offs. That temptation, in the case of the philosophy of science, I shall try to resist, but not to the degree of dismissing its
claims as wholly groundless.

To the question: 'Why science?', there is at least a stock answer, the
Baconian answer: 'Science helps us to understand why things behave as they
do and through coming to understand their behaviour to gain mastery over
them'. In calling this the 'stock answer', I by no means wish to imply that it
is universally accepted. From Plato on, critics of science have argued that
science does not in fact help us to understand *why*, but only *how*, things
happen. Scientists themselves have sometimes said as much. Scientists have
often denied, too, that, except as part of the process of experimenting, the
mastery of things is any concern of the scientist, as distinct from the technologist. Setting aside, as a mere evasion, this disclaimer of any interest in
practical consequences, the enemies of science for their part have alleged
that, far from enabling man to master nature, what science really does is to
increase man's power over man or, even more harshly, that it spins nature
out of control, unleashing forces which man cannot master and which will in
the end destroy not only his entire species but every living thing.[2] If, none
the less, I call the Baconian answer the 'stock' answer, that is because it is
the sort of answer which scientists ordinarily give when they are asked to
justify the teaching of science in schools, or to win public support for scientific research. It is also the answer which critics of science usually take as
their point of departure.

There is, in contrast, no such stock answer to the question: 'Why philosophy of science?' I am going to suggest that one reason why this is so is that
'philosophy of science' is the name given to a very loosely related family of
activities, linked by the fact that they are all philosophical and all related to
science, but belonging to different branches of philosophy and very differently
related to science. That explains, amongst other things, why philosophy of
science is sometimes assumed to be a theory *about* science and sometimes
asserted to be part of science, as when Hilary Putnam writes that 'the philosophy of physics is continuous with physics itself'.[3] For, I shall be maintaining,
certain kinds of philosophy of science are *about* science, whereas other kinds
of philosophy of science form *part* of science. Furthermore, I shall also be

arguing, different kinds of philosophy of science are about, or are parts of, science in rather different manners. As a consequence of this complexity, there is no single answer to such questions as 'Should science students be taught philosophy of science?' One is forced to reply, 'It all depends on what you mean by the philosophy of science', however little one likes that evasive-sounding type of answer.

I shall begin from that sort of philosophy of science which can most plausibly be regarded as forming part of science, which certainly ought to be engaged in only by philosophers who are able seriously to work at science, and which can properly attract the attention of scientists without any feeling that they have moved outside their professional field, that they are taking a holiday into philosophy. I shall call it 'substantive' philosophy of science, for it concerns itself with substantive problems in science.

At a certain point in time, with Newton, physics broke away from philo-sophy, and was transformed into a strictly professionalised discipline. But that break was not an absolute one, as is brought out by the fact that even afterwards it for long seemed appropriate to call physics 'natural philosophy'. In Germany, many philosophers refused to grant science its independence; they continued to construct 'philosophies of nature' in which, with Schelling and Hegel as notorious but far from unique examples, they tried to solve by purely metaphysical analysis questions which, after Newton, were in England wholly left to scientists.

Indeed, hostility to the claims of science to be something more than a bundle of technical tricks is, or until very recently was, extremely common amongst Franco-German philosophers. As Carnap freely admitted, such hostility largely motivated the logical positivist counter-attack on philosophy. In England the situation was very different. There was little overt hostility to science, provided it kept to its proper place, which was, very firmly, downstairs. At the same time, philosophers were markedly reluctant to con-cede either that science had anything to say of which philosophy had to take account or, on the other side, that philosophers could in any way contribute to science. During the hey-day of ordinary language philosophy, only a very few 'outsiders' — most conspicuously Reichenbach — paid any heed to those problems which lie, to take over the title of J. J. C. Smart's book, 'between science and philosophy'. But more recently, as the debilitating effects of ordinary language philosophy have worn off, there has been a spate of writings on such topics as Space and Time, in Australia for example from J. J. C. Smart, Ian Hinckfuss, Graham Nerlich.

Some of this work has grown out of what is widely felt to be the unsatis-

factory theoretical condition of contemporary physics and an accompanying
conviction that the grounds for dissatisfaction are unlikely to be removed
either by fresh experimental work or by purely mathematical innovations.
It is sometimes argued, indeed, that quantum mechanics and relativity theory
have brought philosophy back into theoretical physics, demonstrating that
the ideal of a physics which would be wholly independent of philosophy was
a quite unrealisable one. Quantum mechanics and relativity theory, so it
is then said, make use of epistemological assumptions, in so far as they
necessarily introduce into physics the concept of a *perceiving person*, and
automatically raise, therefore, the traditional philosophical problems about
the relationship between the perceiving person and what he perceives.

This interpretation of the present situation in physics as necessarily involving a reference to a perceiver is by no means universally accepted. Karl
Popper, for one, has attacked it, but in a manner highly germane to the
present argument. He agrees that a subjectivist epistemology has penetrated
not only quantum mechanics but also the theory of probability, statistical
mechanics, the theory of entropy and information theory.[4] But this, he says,
is to their detriment. They do not need such an epistemology; what they
really call for, he has sought to show, is an 'epistemology without a knowing
subject', an epistemology of the sort adumbrated by Plato, Hegel, Bolzano
and Frege. If Popper is right, then a considerable body of recent scientific
theory is vitiated by its adherence to an inadequate epistemology – as distinct
from either experimental or mathematical errors. So Popper still supposes that
philosophy has a positive role to play within physics, that to get its theories
right physics will have to reconsider philosophical issues. Many of those who
reject Popper's philosophical views would at least agree, as Hilary Putnam
does, that 'no satisfactory interpretation of quantum mechanics exists
today'[5] and that philosophy can help in finding one. Carnap suggested that
if present-day theoretical physics is to assume an intellectually more satisfactory form, this can only be as a consequence of 'close co-operation
between physicists and logicians'.[6] Mario Bunge was prepared to write,

I ... think that it is the philosopher's duty to remind the scientist that most of his
achievements are bound to be provisional, and that it is the philosopher's privilege to
speculate on possible solutions to problems that are not solved in a satisfactory manner
or that have not even been noticed by scientists – as long as the speculator proceeds
knowledgeably and imaginatively, and that [sic] he is willing to listen to scientific
criticism.[7]

Now, if philosophy of physics overlaps in this way with physics – and a case
can certainly be made out for saying something similar of biology in respect

to disputes about, for example, 'species' or 'genetic resemblance' or 'evolutionary taxonomy'[8] — then the point of one kind of philosophy of science is simply identical with the point of science. That such philosophy of science is contributed to by philosophically-minded scientists — Heisenberg, for example — as well as by scientifically-minded philosophers strengthens rather than weakens the case for saying this. There is no question of philosophy coming in from outside as a doctor to science, in the manner Watson condemns. Scientists, on the view we are considering, have themselves been forced into philosophical argument and forced to make use of philosophical premises, as part of doing science; these premises, these arguments, are properly subject to criticism by philosophers just as the scientists' mathematical premises and mathematical reasonings are properly subject to criticism by mathematicians. Of course, one might try to avoid such conclusions by arguing, let us say, that quantum mechanics is nothing but a set of practically-useful devices, or nothing but a set of equations. It is worth noting that Putnam speaks of the unsatisfactory 'interpretation' of quantum mechanics, as if quantum mechanics were one thing, its interpretation quite another thing, as if, that is, quantum mechanics proper consists simply of equations. So if might seem that philosophy of science of the sort I have been discussing is after all *about* science, about its interpretation, and not a part of science. But if we are not content to think of physics either as mathematics or as a set of practical devices, if we suppose that it tells us something about the world, then its equations *have* to be interpreted. The interpretation cannot be dismissed, that is, as inconsequential gas, no part of 'the real stuff' of the theory; on the contrary, it forms part of science. 'You can't do science in full self-conscious understanding', so Lawrence Sklar has argued, 'unless you realise how much it depends upon philosophical modes of reasoning as well'.[9] If this be so, then any teacher who hopes his pupils will do science with a 'full self-conscious understanding' will have to introduce them to substantive philosophy of science.

A rather different kind of philosophy of science — the distinctions between members of this family of inquiries are not hard-edged — arises in opposition to the view, often propounded by scientists, that conclusions which have traditionally been supposed to belong to metaphysics and to be defensible only by the characteristic types of philosophical reasoning are in fact deducible from well-supported scientific theories or falsified by specific scientific experiments. No philosopher, it would be generally agreed, can now write quite as Kant did about the geometry of Space, not after non-Euclidean geometry and Einstein's theory of relativity. But some philosophers of science still question whether in this area it directly follows, without the help of

any purely philosophical assumptions, that there is no such thing as Absolute
Space or, at the very least, as Absolute Space-Time.

They are even more sceptical about such claims as that science has now
shown once and for all that the doctrine of representative perception must
be true, or that determinism must be false, or that mind must be distinct
from body, or that Berkeley's idealism is substantially correct, or (more
recently and on the wilder shores of California) that the mystics were right.
We are no longer dealing with a situation, as we were in discussing substantive
philosophy of science, where the scientist consciously makes use of philosophical premises. Rather, he is professing to derive metaphysical conclusions
without the aid of philosophical premises or — as also happens — ethical conclusions without the aid of normative premises. And if he can really do this
then there is no good reason for denying that these propositions now form
part of science, that they count amongst scientific conclusions. So Sir John
Eccles, for example, takes it to be a scientific conclusion that the mind is
something other than the brain.

Such claims generate what I shall call 'critical' philosophy of science, the
immediate aim of which is to demonstrate that these metaphysical or ethical
propositions *do not* form part of science, are not immediately deducible from
scientific premises. It is a very important question how far science can carry
us. We quite commonly encounter the pronouncement: 'Science has shown
that . . . ', where what is allegedly shown is a metaphysical, or an ethical,
conclusion. A familiar example is the attempt, by no means dead, to derive
ethical conclusions from Darwinism. Philosophers can do philosophy a great
deal of damage by ignoring science when it is relevant to their concerns, or
by defensively denying that it *can* be relevant. But so can scientists do a great
deal of damage, not only to science but to human culture generally, if they
wrongly presume that they can 'prove scientifically' conclusions which follow
only on certain philosophical presuppositions. In so far as critical philosophy
of science can generate a sense of not only the interplay, but also the difference, between philosophical and scientific arguments, it can help to confine
scientific *hubris*.

Much of the argument by which philosophically-minded scientists profess
to derive their metaphysical or ethical conclusions from scientific premises is
so naive that philosophers do not trouble themselves to analyse it critically.
Few philosophers bother themselves with Teilhard de Chardin or even with
Jacques Monod. But one familiar example of critical philosophy of science is
Susan Stebbing's *Philosophy and the Physicists* (London, 1937) in which she
set out to show that the philosophical conclusions which Eddington and Jeans

professed to have demonstrated did not follow from the evidence which purported to establish them, that they therefore did not form part of science, and that the argument which professed to show that they did displayed certain philosophical weaknesses.

Critical philosophy of science lies on the controversial boundary between philosophy of science which forms part of science and philosophy of science which is about science. From the point of view of an Eccles, any argument about the relationship between the mental and the physical forms part of science; Australian materialism, let us say, is simply bad science. But those who reject the view that it is possible, by purely physiological investigations and without any appeal to philosophical premises, to demonstrate that minds are immaterial objects — as distinct from arguing that in some particular case this has not been done — are committed to a view about science, about what it can, or cannot, establish from its own resources. At this point, critical philosophy of science turns into what I shall later call 'co-ordinative' philosophy of science.

To turn now to philosophies of science that are less controversially *about* science, it is nevertheless by no means uncontroversial in what sense they are 'about' it. Perhaps we should begin by trying the effect of lifting the stock Baconian answer about the point of science to a meta-level. Philosophy of science, we should then say, sets out to understand why *scientists* (as distinct from natural phenomena in general) behave as they do, and through such an understanding to gain control over their activities. This is not an answer which would appeal to most Anglo-American philosophers of science. They would tend to reply, first, that understanding why scientists behave as they do falls within the province of psychology or sociology and has nothing to do with philosophy and, secondly, that philosophers have neither the desire nor capacity to control the activities of scientists, that indeed scientists should be left free, *not* controlled.[10] But the contrary view is widely held by Marxist philosophers of science — is held, that is, over a large part of the surface of the globe. Science, on their view, is now one of the major factors of production, to be set alongside the traditional land, labour, and capital; we need to understand it in order to govern its contributions to the economic and social order. A philosophy of science is to be adjudged, as they see it, in two ways; first, by its 'adequacy to developing science' and, secondly, by 'its connection with progressive social practice'.[11] It must help us, that is, to understand the direction of scientific progress at any particular time and to see how that progress can be applied for social benefits. Now, of course, in Marxist countries this conception of the philosophy of science is entangled with, first, a dialectical historicism according to which no general description can be applied to science, as distinct from a description of how it is moving at a given time and, secondly, a political thesis that capitalism, unlike socialism, neither gives science and technology the attention which they deserve as productive forces nor is capable of coping with the social changes they create. But we should not lightly conclude that except on these assumptions philosophy has nothing to say on such topics, even if we deny that they constitute the central, let alone the sole, concern of the philosopher of science. There are questions here which can properly be discussed within what I shall call *social* philosophy of science.

Up to a point, however, the Anglo-American philosopher of science is clearly right. There are a great many inquiries into science, a great many doctrines about science, which do not fall within the province of philosophy. Whether science is parallel. as a factor in production, to land, labour and capital, is a question for political economists, not for philosophers. The role of age in determining the chance of making fundamental scientific discoveries, whether great scientists tend to be first-born children — these are questions for psychologists. The study of the institutions through which society subsidises scientific activity lies within the province of political science. Not every inquiry which is about science, in other words, is philosophical in character. We still have to discover what constitutes a *philosophical* inquiry about, as distinct from within, science.

Attempting to reply to this question, we should first note an ambiguity attaching to the word 'science' — as indeed it does to 'art' or to 'philosophy'. By 'science', we sometimes mean a form of human activity. To seek research funds for science is to ask that scientific activities be supported. In such phrases as 'a science-based technology', on the other hand, 'science' means the formulated outcome of scientific activities. 'Learning science' is therefore also ambiguous, in an educationally important way; it can mean learning *to be a scientist* or it can mean learning *what scientists have discovered*. 'Understanding science', similarly, can consist either in understanding scientific propositions or in understanding what, to put it colloquially, 'makes scientists tick' — and not in a sense which is proper to psychology. Again, when it is said that science teachers should emphasise 'the structure of science', that is sometimes taken to mean that they should avoid detail and concentrate on the major scientific concepts, and sometimes that they should embark on the philosophy of science, in the manner of Ernest Nagel's *The Structure of Science*. And the two are often amalgamated, as if they were identical.[12]

Of course, this ambiguity is not an accidental one. The object of the scientist is to make a contribution to formulated science, even if the journey often interests the scientist more than does the arrival. But formulated science is only a small selection out of the results of scientific activity, many of the results of which are literally or metaphorically poured down the drain. What goes into a text-book is a *very* small selection, consisting only of exceptionally successful science. And that is all the science most students will ever get to know or many technologists will ever use.

Can one say that formulated science simply describes the activities of the scientist? Clearly not. A published article is a very stylised account of what a scientist does. It doesn't include such sentences as 'on Tuesday, I dropped a flask and had to start again' or 'on Monday, waiting at the traffic-lights, I had a bright idea'. Such books as *The Double Helix* or Faraday's diaries may come nearer to describing the activities of scientists. But we do not think of these as formulated science. Formulated science asserts and justifies conclusions; it does not attempt to describe how they were actually arrived at.

Now, on some accounts of the matter — indeed, this has been by far the commonest view — the philosophy of science is entirely about formulated science. Its interest lies, as the point is sometimes put, in 'the context of justification', not in 'the context of discovery'. On the most naive interpretation, still wide-spread in the public at large, formulated science consists of two elements: (a) solid evidence, (b) conclusions derived from that evidence. Philosophy of science, on this view, is about the formal relationship between (a) and (b). Not in detail, of course. The philosopher of science does not work through a text-book on chemistry with a sort of formal slide-rule and tick off the scientific conclusions which are adequately supported, crossing out those which are not. His argument remains at a very general level. But he does, on the view we are considering, look for algorithms which could, in principle, be used in this way — at least if the text-book fully set out the evidence, instead of relying on illustrations. I shall call such a philosophy of science 'logistic' philosophy of science.

There is always a tendency for 'philosophies of' to collapse into philosophy — for books on the philosophy of education, let us say, to be indistinguishable, except for an example or so, from books on moral or political philosophy. In the case of logistic philosophy of science, this can happen with particular ease. Scientific evidence and scientific conclusions are often defined in purely formal terms: conclusions as universal propositions (including statistical propositions under that head), 'evidence' as singular propositions. Then the question whether scientists are justified in their conclusions is identical with the question whether any of us, scientist or not, is *ever* justified in accepting as true a proposition which is universal in form on the evidence of particular cases, in accepting as true such a proposition as, to take a favourite example from writings on the philosophy of science, 'all ravens are black', on the ground that all the ravens we or other people have encountered have been black.

That in discussing that issue the philosopher is raising a general logical problem, not a problem specific to the philosophy of science, comes out in the mere fact that this example — or its predecessor 'all swans are white' — is taken to be typical. For what it is typical of is a pre-scientific generalisation. It differs from scientific generalisations (at least in what I have elsewhere called 'aristoscience'[13]) in a great many, closely connected, ways:

(1) We have no *theoretical* grounds for believing that all ravens are black. (The theory of natural selection does not supply such a reason; it is quite compatible with a grey raven's turning up in some remote New Guinea valley, where being of a lighter colour than black confers survival advantages.)

(2) 'All ravens are black' is a strikingly *independent* proposition in the sense that if a grey raven did turn up, we should have to revise very few other propositions except such as are derivable from it by purely logical means. The discovery of black swans upset the poet but not the scientist.

(3) The 'evidence' for it is singularly direct. No auxiliary propositions of a theoretical sort intervene between it and such evidence as 'this is both a raven and is black'.

(4) It is about familiar objects and familiar properties, which any normal child can readily be taught to recognise, without benefit either of scientific theories or scientific equipment.

If we go on to ask how 'all ravens are black' is related to those propositions which count as evidence for it, there do not seem to be more than a very limited number of possible answers, such answers as, to put them in their crudest form:

(1) The inductive answer. Propositions of the form 'this is a raven', 'that is a raven', . . . together entail 'all ravens are black'.

(2) The probability-calculus answer. Although propositions of the form 'this is a raven', 'that is a raven' do not together entail 'all ravens are black', they do entail that the proposition 'all ravens are black' has a certain probability which is very high if the number of instances is very large.

(3) The anti-logistic answer. The set of propositions 'this is a raven', 'that is a raven' neither together entail nor, however large in number, make highly probable the proposition that 'all ravens are black'.

As I have admitted, the three possible lines of argument here distinguished are very crudely expressed. (I have entirely ignored the question Hempel raised, whether 'this is not a raven and is not black' counts as evidence for 'all ravens are black'.) It is sometimes argued, for example, that the only important question is whether we have logically well-founded reasons for believing that the next raven we see will be black, that we have no need of such general propositions as 'all ravens are black', whether inside or outside science. Or again, that the role of evidence is to *confirm* generalisations, not to provide data from which they are derived. Or again that what really matters is that no non-black raven has yet turned up. But let us assume, as perhaps most although by no means all philosophers of science now do, that such attempts to avoid the anti-logistic answer fail, and that in the end they are all reducible to induction. Let us also make the large assumption that there is no inductive logic which will enable us to make the step from 'this is a black raven', 'that is a black raven', whether to 'all ravens are black' or to 'the next raven will be black'. Let us assume, too, that when the generalisation is infinite in its range, as physical laws are, its probability on such evidence will be zero, and that even when its range is merely very large, as is true of almost all scientific propositions, its probability will be extremely small. Let us assume, in other words, that although scientists will of course make use of deductive logic in their reasonings and of the probability calculus whenever they have need of it, there is no 'inductive logic' which will enable them to derive true general conclusions by valid reasoning from 'evidence' and no special 'logic of confirmation' which will enable them, after testing a hypothetical generalisation against evidence, formally to conclude that it must be true or at least highly probable. Or if these assumptions are boggled at as far too sweeping, let us make the more modest assumption that, even if such a logic could be used to derive such pre-scientific generalisations as 'all ravens are black', there is no way of working with an inductive logic in the actual context of science, where complex auxiliary theories play so large a part in any testing process.

These assumptions will obviously affect the value of the philosophy of science. It would be very pleasant to be able to say of any hypothesis *h* that given the body of evidence *e*, it must be true or must at least be highly probable — especially if *e*, as in most classical inductive logics, is taken to consist of theory-free observation statements, about the truth of which there can be no serious doubt. One need not be at all surprised that many philosophers of science are still intent on working out a way of enabling us to do this. No one could have any qualms about the value of a philosophy of science which provided science with a logic of this sort, even if, as I have suggested, it would properly speaking be a general logic, exactly as deductive logic is a general logic, applying to all generalisations and not only to scientific generalisations. It would form part of the philosophy of science rather than of logic only if one could show that such a logic works in science but not outside it, e.g. that only in science does one begin from really firm observation statements.

But suppose we accept the anti-logistic view. Suppose we deny that there is any such thing as 'inductive logic' or 'a logic of confirmation'. Are we committed to the conclusion that philosophy of science has nothing of value to say about scientific reasoning? Well, in the first place, one should not dismiss as practically valueless the anti-logistic answer. What it amounts to is this: no evidence is sufficient to establish by formal means that an open universal generalisation — for simplicity's sake I shall have to leave aside the very important but still more complicated case of statistical generalisation — is either true or is very probable. And that conclusion can help to 'loosen up' science, to prevent it from collapsing into a dogmatic rigidity; it encourages speculation, reconsideration, an alertness to the possibility that this or that 'established scientific truth' may in fact need to be questioned, open-mindedness about alternative possibilities.

But on the other side, the reply might come, it encourages *scepticism*, and that is even more fatal to science than is dogmatic rigidity. It leads in the long run to Feyerabend's later position that there is no better reason for paying attention to what scientists tell us than there is to what seers and prophets tell us. It opens the gates, in other words, to anarchic irrationalism. Proceeding with any specific inquiry, this argument would continue, the scientist needs to feel that he can rely upon background knowledge, that he is not moving from quicksand through quicksand into quicksand. Science is sustained as an institution only by the scientist's belief that he is in pursuit of the truth, and has at least some prospect of attaining to it. As for technology, it looks to science for reliable knowledge, not for mere speculation. And only an inductive logic, the conclusion then runs, can justify that confidence. Popper, in rejecting such a logic, was the thin edge of a sceptical wedge that threatens to split science asunder.

From Plato's *Theaetetus* onward, philosophers have been obsessed by the need to find some way of replying to the sceptic. Scientists, with a few unusually philosophical exceptions like Descartes, have never taken scepticism seriously. True enough, nineteenth-century agnosticism was largely propounded by scientists. But agnosticism was not scepticism about science, it was scepticism about metaphysics. If philosophy of science issues in scepticism then it is impossible to rebut the condemnation of it as 'morbid'. So long as neither scientist nor technologist starts complaining that when he goes to formulated science for reliable information it constantly lets him down, no scientist is going to take scepticism seriously.

But is anti-logicism sceptical? Many philosophers have set out to show that it is not, with the aid of what I shall call *epistemic* philosophy of science. Without themselves putting the matter quite in this way, they have sought to show that it is with good reason that technologists turn to science, even though the scientist's evidence is not related to the scientist's conclusions in the manner supposed by any form of logistic philosophy of science. The philosopher of science cannot, to be sure, provide the scientist with a method of demonstrating that any particular theory is correct. Indeed, the possibility always remains that it is *not* correct. But he can provide him with good grounds for taking theory t_1 to be better theory than theory t_2. I shall consider as my leading example Karl Popper.[14]

One is at first inclined to say that Popper talks about scientists rather than about formulated science. But certainly he is not attempting to describe what scientists *actually do*. 'I am inclined to say', he indeed writes, 'that we should attempt to find out what they "ought" to do'. This, he explains, is the 'ought' of a *hypothetical* imperative. 'The question is', he continues, 'how should we proceed if we wish to contribute to the growth of scientific knowledge?' And his answer runs thus: 'You cannot do better than proceed by the critical method of trial (conjecture) and the elimination of error, by trying to test, or refute, your conjectures'. 'The argument supporting this reply', he continues, 'really belongs to situational logic. I do not think we should turn to the (sociological) question of what scientists do or say, except perhaps to refute certain competing answers'.[15]

It will be noted that Popper describes the arguments in favour of his view as belonging to 'situational logic'. One must not be misled by his use of the word 'logic' into supposing that his is just another form of logistic philosophy of science, with the peculiarity that he relates universal generalisations to potential falsifiers rather than to confirming instances. Situational logic, for Popper, concerns itself with the rational procedures to adopt when we are confronted with a class of problems; it issues in methodological rules, not in algorithms. The logistic approach, as he sees it, is misguided because it begins from 'evidence'. The so-called 'evidence', on Popper's view, is nothing but a set of well tested hypotheses which, although they are in no sense indubitable, scientists accept as 'basic' and use in order to test theories. (Some of his critics would immediately conclude that Popper is, in consequence, sceptical; he abandons that search for an indubitable starting-point which has inspired a

great deal of philosophical reflection.) Science begins not from this 'evidence',
says Popper, but from problems; its concern is with the growth of knowledge
through the process of severely testing broad hypotheses and in the process
replacing propositions with lesser verisimilitude by propositions of greater
verisimilitude. We must not suppose that what survives after this process of
testing – a process which Popper compares to Darwinian natural selection – is
'reliable knowledge', if we mean by this a piece of information (in a very broad
sense of 'information') which it is safe to rely upon *absolutely*. All we can
properly say is that it has so far stood up to vigorous attempts to refute it, or,
at best, that it has greater verisimilitude than its rivals. So, faced with a situation in which we have to act, it is more rational to rely on formulated science
than it is on alternative sources of information which have not been subjected
to such testing. (Very roughly, this is like relying on information in the *Encyclopedia Britannica* rather than on information in some popular newspaper;
the first will have been checked much more carefully than the second. All the
same, the *Encyclopedia Britannica* may be wrong, the newspaper article right.)

The fundamental question for epistemic philosophy of science is whether
this answer suffices to dispel scepticism. But we can also ask how one could
test an epistemic philosophy of science. Popper tells us that he is setting up a
methodological rule, which is therefore not refutable by reference to the
actual practice of scientists. If he were issuing a categorical imperative, that
would be true; scientists who do not proceed in the manner Popper prescribes
– those, for example, whom Kuhn calls 'normal scientists' – would, in terms
of that imperative, be bad scientists. But then we should naturally wonder
what gives Popper the right categorically to assert that scientists ought to do
this or that. We remember Watson's remarks about the philosophical doctor –
a doctor with no real credentials – who steps in to tell scientists what they
ought to be doing. In fact, however, Popper's imperative is supposed to be
hypothetical. He is saying 'if you want knowledge to grow, keep to my rules'.
But this advice rests on the more general assertion: 'knowledge grows only
when scientists proceed in the manner I recommend'. And any such assertion
does have to meet the objection, which can only be drawn from the history
of science, that these are *not* in fact the procedures which lead to the growth
of knowledge, or that they mark only rare and quite untypical examples of
scientific progress.

Of course, the matter is not so straightforward as these remarks might
suggest. 'History', one might argue, is in the same position as any other
evidence. It consists of tested hypotheses about what happened. No more
than any other evidence can it be theory-free; in particular it is influenced by

the methodological theories of the historian. The historian of science, it might therefore be concluded, tells the story of what would have happened had the scientist acted in accordance with these theories, just as, so it is sometimes suggested, the general historian describes what would have happened had the participants acted in accordance with what *he* regards as rational principles. History, that is, is a *rational reconstruction*; with a different concept of rationality, history will be written differently. Hence historical evidence cannot be used as a way of refuting a methodological hypothesis. Only history written by Popperians could be so used. But then it would be so written as not to refute Popper.

Some of you will recognise here an extended version of the 'incommensurability' thesis developed by Kuhn and Feyerabend, according to which evidence cannot be used to choose between very general theories, since the so-called 'evidence' will already incorporate one or the other of such theories. But, in the present instance, how are we to suppose that epistemic philosophy of science ever gets going? It arises, on Popper's view, out of a problem, not out of a metaphysical intuition of the essence of scientific rationality. And the problem can only be to determine what constitutes the difference between progressive and unprogressive stages in the history of science. (Not, of course, *under what social conditions* science flourishes, that forms no part of an epistemic philosophy of science.) This history of science is written before the epistemic theory which arises out of reflection on it; if it is written from the standpoint of a theory, this is not the theory it is being used to defend. Of course, the epistemic theory can then stimulate a reconsideration of the history of science — this is a typical example of what is now often called a 'hermeneutic circle'. But unless it is to be a mere thought-experiment masquerading as history, any such re-writing will have its boundaries determined by the records — records made, no doubt, by people with theories, but not necessarily with methodological theories. Only by reference to such records is there any way of choosing between Popper's view that knowledge grows by conjectures and refutations, and such alternative views as that it progresses by the accumulation of data or, as Feyerabend suggests, by adopting a policy of 'anything goes'. 'Given any rule, however fundamental and necessary for science', Feyerabend writes, 'there are always circumstances when it is not only rational to ignore the rule, but to adopt its opposite'. 'This', he adds, 'is not just a *fact* of the history of science. It is both reasonable and absolutely necessary for the growth of knowledge'.[16] Unless this is indeed 'a fact of the history of science', however, it is a mere *a priori* dictum, to which we need pay not the slightest attention.

However it be established, there can be no doubt, I should suggest, about the point of epistemic philosophy of science. If there are ways of proceeding which are essential to the growth of knowledge, which constitute scientific rationality, which distinguish science from other less successful attempts to secure knowledge, then this is a fundamental fact, with which the scientist ought to be acquainted. Even if 'anything goes', he ought to know this, too, to prevent him from succumbing to unduly confining conceptions of what counts as 'proper science'. He would certainly be very glad to learn from Lakatos when to detach himself from a declining research programme in order to attach himself to a progressive research programme — if Lakatos could tell him. Even if, as Feigl [17] maintains and Feyerabend denies, all such considerations belong to the 'context of discovery' and not to the 'context of justification', there is no good reason for claiming that a philosophy of science is interested only in *justification*, not in *discovery*. To assume as much is tacitly to identify philosophy of science with *logistic* philosophy of science. The practical problem for the science teacher, of course, is that he is confronted not by a single epistemic philosophy of science but by competing epistemic philosophies of science, as even our schematic survey of a few contending alternatives will make plain. When J. W. Mellor wrote his textbook on inorganic chemistry, he could begin with a chapter on what he took to be an accepted view about the epistemology of science. That is no longer true, if it ever was. But once again the very diversity of opinions may 'loosen up' the too rigid professionalisation of science.

In a certain sense, both logistic and epistemic philosophies of science try to justify science, to show that science is rational in a sense in which, let us say, occultism is not, that it accepts or rejects theories for good reasons. They differ about the character of these reasons, but not about the need to show that science proceeds in a manner which justifies its reputation. In contrast, what I shall call 'structural' philosophy of science, exemplified in Ernest Nagel's *The Structure of Science*, begins from the presumption that science is a successful going concern. Its concern is to analyse what counts within science as a sound explanation, a useful model, a suggestive analogy, a successful theory, a good definition or classification. 'The conclusions of science', so Nagel argues, 'are the products of enquiries conducted in accordance with a definite policy for obtaining and assessing evidence'. But, Nagel goes on to concede, 'it must be admitted that the canons for assessing evidence which define the policy have, at best, been explicitly codified only in part, and operate in the main only as intellectual habits manifested by competent investigators in the conduct of their inquiry'.[18] He rejects, that is — if with some signs of regret — the logistic claim to be able to justify formally the scientist's 'intellectual habits'. Nevertheless, he goes on to say, the success of this policy leaves 'little room for serious doubt concerning the superiority of the policy over alternatives to it'. As I began by suggesting, that is, structural philosophy of science begins from the assumption that science is a successful going concern, indeed the *most* successful of intellectual going concerns. The object of science, it also presumes, is to offer explanations by reference to facts, good theories, principles, laws; the formal logic it uses is deductive logic. Beyond this, what Harré in a similar vein calls *The Logic of Science*, consists of successful strategies.

This sort of philosophy of science attempts, then, to 'raise the scientist's consciousness' about the strategies which as a scientist he will use, and to help him to distinguish a proper from an improper use of these strategies. The practising scientist may find it somewhat too skeletonised; there is not much of the hot blood of science in it. On the other hand it is, at least, one of the things which writers on the teaching of science have in mind when they assert that, at school and university, scientists should be made acquainted with the structure of science. From my own relatively slight acquaintance with teaching a structural-style philosophy of science to university science students, I believe that it can be a distinct stimulus to intelligent scientists. But its quasi-pragmatic approach will leave many philosophers distinctly unhappy, as it certainly leaves Habermas unhappy.[19]

Cutting across the distinctions we have recently been making there lies 'ontological' philosophy of science. It arises very naturally from any attempt to approach the philosophy of science by way of explanations, in so far as such explanations very often refer to entities which lie outside the range of everyday observation. So it is not surprising to find that Nagel, with his emphasis on explanation, raises questions which lie in its sphere. But it can also have other sources.

Ontological questions can be asked about any proposition. So one can ask what we are ontologically committed to when we assert that 'all ravens are black', whether this presupposes that there are particular entities, ravens, and a property of being black, or whether in its proper logical form it is about relations between properties — 'if anything has the property of being a raven, it has the property of being black' — or whether what it is 'really about' is a set of relationships between sense-data. Such questions scientists would normally set aside, as no business of theirs. And quite properly.

They can less readily do so, on the face of it, when they make assertions about, let us say, neutrinos. Is the physicist committed by making such assertions, the ontological philosopher of science asks, to asserting that neutrinos are as much part of the furniture of the world as are ravens? Does the scientist *discover* neutrinos in the same sense in which a field naturalist might discover a new species of animal, or does he invent them somewhat as Dickens invented Mr. Pickwick, if with a very different purpose?

These questions inevitably lead to a similar set of questions about the nature and the role of scientific theories, whether such theories offer us explanations of the way things behave as they do, or, as Mach thought, a compendious description of their behaviour, or, as the instrumentalists argue, a mere device for getting from one set of experimental observations to another. And, linked with this, whether it is appropriate to apply to scientific theories such predicates as 'true' and 'false' or only such predicates as 'convenient', 'simple', 'useful'. One can ask, too, about the relations between theories in different areas of science, for example whether biological theories are reducible to theories in physics and chemistry, and what ontological consequences, if any, follow from a negative or a positive answer to this question.

No doubt, the scientist might reply that none of this matters, that he need not commit himself, on such topics, to one view or another, that he can go ahead with his work indifferent to all such ontological controversies, that it makes no difference whether he is a 'realist', or an 'instrumentalist', or a 'phenomenalist', or what you will. But I doubt whether this is true. If one thinks of a neutrino as nothing more than a mathematical device, then there would be no point in setting up experiments to look for neutrinos, if one thinks of theories in an instrumental way, then one may neglect incoherences within science where a determination to resolve these incoherences might give rise to better theories. A 'realist' approach to science, in contrast, encourages attempts to look for as-yet unobserved entities and to remove incoherences. It is at least plausibly argued by Karl Popper that the Copenhagen complementarity theory arose out of an instrumentalist conception of scientific theories and has handicapped the subsequent development of theoretical physics by treating an incoherence as unimportant.[20] A scientist is, I should say, deluding himself, if he supposes that he is not *in fact* either an instrumentalist or a realist. And it is better for him to have some consciousness of the issues involved.

Beyond this, instrumentalism, at the hands of teachers, encourages the view that science consists of formulae plus technical tricks. It encourages that scorn for science which many humanistically-minded school children carry away from school science courses. So ontological philosophy of science has both a heuristic and a cultural value. It encourages the scientist to consider what he is trying to do, what tasks it is sensible for him to undertake; and in the process it helps him, and his students, to consider what kind of contribution science makes to man's *understanding* of the world, as distinct from his capacity to find his way around in it by the use of sign-posts.

That remark leads me into another variety of the philosophy of science which, in default of a better name, I shall call 'co-ordinative' philosophy of science. This compares science with, relates it to, and demarcates it from, the work of participants in those other forms of activity which have, on the face of it, some similarity to science. Theology, art, philosophy, technology, mathematics, history, the social sciences, all fall into this category. And so does ordinary commonsense discovery and invention. To take an instance, J. G. Kemeny has gone so far as to assert that 'all of Science is applied mathematics'.[21] It is sufficiently apparent, certainly, that science uses mathematics in a variety of ways — the ways, for example, that Brian Ellis discusses in *Concepts of Measurement*. Yet against a Kemeny-type view, Hanson argues that 'physics is not applied mathematics'. 'It is', he says, 'a natural science in which mathematics can be applied'.[22] Which is right? This is a problem in co-ordinative philosophy of science. As for the relationship between philosophy and science, that is anything but straightforward, much less straightforward than I have tended to assume. All philosophy, some would say, forms part of science.

Broader questions of the same sort arise if we ask whether there can be 'social sciences' and if not, why not. Or whether history can be 'scientific'. Or, for the matter of that, art. The 'placing' of science in human culture has been most systematically explored by Hegel, Collingwood and Whitehead. In a much more modest way I have tried to contribute to it when writing about the philosophy of history, about the grounds for taking art seriously, or in *Science and its Critics*. Perhaps, therefore, I am inclined to exaggerate its importance and foolishly to regret that it has been so largely left to philosophers of an Idealist persuasion. Yet no scientist, surely, can be wholly indifferent to either of two very important cultural movements, scientism, the doctrine that outside science there is no truth except of the most trivial sort, and anti-science, the doctrine that the very methods which science uses disqualify it from arriving at truth as distinct from technical mastery. And both those cultural movements depend on a comparison between science and other forms of human activity, whether these other forms of human activity are taken to be 'inferior' or 'superior'.

Finally, let me briefly refer to what I shall call the 'social' philosophy of science, incorporating moral philosophy within the broad concept of social philosophy. Here I return at last to those Marxists from whom I originally set out. Science was once, and not so very long ago, a gentlemanly pursuit conducted for the intellectual edification of gentlemen. For all the premature prophecies of a Bacon and a Descartes, not until the last decades of the nineteenth century did it become plain that science also was, or could be, a factor in production, a social force; that, indeed, its presence in or absence from a society very largely determined the prosperity of that society. Both in England and in France, scientists were for long reluctant to admit as much. Germany became a European leader partly because it was the first country fully to recognise the new social possibilities science had created.

For some time thereafter, it was widely if not unanimously believed that the realisation of these possibilities could only be beneficial. Indeed, science was presented to the world as a moral example. Scientists possessed all the major virtues — industry, imaginativeness, disinterestedness, objectivity, honesty, cooperativeness — and yet in a form which a Utilitarian could applaud as conducive to the greatest happiness of the greatest number. At last the moral ambitions of a Shaftesbury were satisfied, the cynicism of a Mandeville dissipated. Private virtues were, conspicuously and directly, public benefits.

That euphoria, alas, did not last. There were always, of course, those who feared science. Not only philosophers and theologians, who saw in it a threat to their world-view, but artists and novelists. Mary Shelley wrote her *Frankenstein* as early as 1818. But now moral and social issues are much more sharply and widely raised, not only about the outcome of science but even about the attitude of mind which science engenders. A typical question is this one: 'Can a scientist who knows quite well that anything he discovers in the area in which he is working will be transformed into a weapon of destruction nevertheless disclaim all responsibility for the creation of such weapons?' Many scientists would reply in the affirmative, disclaiming all responsibility for what they call 'the use to which their discoveries is put'. But is this a morally tenable view? Such questions can scarcely be set aside as of no importance. All that has happened, one might nevertheless suggest, is that moral philosophy is now critically *applied* to science. On this view, one can no more properly talk of a special social (or moral) philosophy of science than one can of a special social (or moral) philosophy of business. But the situation here is much what it is in other branches of the philosophy of science. Science generates a set of moral problems, just as it generates a set of epistemological problems, which we would not otherwise have to confront. Moral philosophy of science discusses this set of problems. And a scientist ought not simply to ignore them. In particular, the dilemma in which many of us now find ourselves, in which our traditional adherence to the principle of freedom of inquiry has to confront a growing uncertainty about the outcome of scientific inquiry, can scarcely be set aside, either by scientists or by philosophers, as insignificant.

More generally, the social philosopher of science can ask whether science in fact exemplifies, as has so often been supposed, the classical radical values of liberty, equality and fraternity — liberty as liberty to enquire, equality before the tribunal of science, and fraternity through membership of an international brotherhood of scientists. This, some of its critics would say, is a gross Romanticisation of the present condition of science. It might roughly apply, they will perhaps grant, to 'the golden age of science', the nineteenth century, but it bears as little relation to contemporary science, the science of research teams, as does the Romantic description of the mediaeval craftsman to the modern industrial worker. In fact, they would say, science is now stringently controlled by secrecy restrictions and, what is even worse, confined within professional orthodoxies which narrowly limit the scientist's freedom of inquiry. It is intensely hierarchical, with its rigid distinctions between directors, principal inquirers, research assistants and laboratory attendants. Far from being fraternal, it is bitterly competitive, often along nationalistic lines. The laboratory, they might further argue, is but a special sort of factory in which the investor — state or private — buys labour to create useful devices, a 'theory' being but one such device. But do not such critics overlook, or set aside as inconsequential, what still distinguishes scientists, namely their passion for discovery? That is the sort of question social philosophy of science attempts to explore.

To sum up, I have distinguished, very briefly and inadequately, a number of different kinds of investigation which can all be brought under the umbrella of 'philosophy of science'. This is not at all in order to suggest that they are sharply sundered and ought to be the subject of rigorously defined specialties. Philosophy, in general, is not like that; the moral philosopher finds himself driven into ontology or epistemology and his conclusions can present problems to the logician. But I have wanted to emphasise that philosophy of science incorporates a wide range of philosophical inquires, some of them closely associated, others more loosely linked, some of them mutually consistent, others not so; that there is more than one point, then, at which philosophy bears upon science. Most books on the 'philosophy of science' concentrate on only one, or a limited sub-set, of these problems.

In a necessarily summary and dogmatic fashion, I have also asked how far these questions ought to be taken seriously by scientists as distinct from philosophers. That still leaves almost everything to be said about the pedagogy of science teaching. But I have assumed that no science student should leave school and university without having some sense of science as a living, active process of inquiry, not just as a set of formulae and professional tricks. And with some sense, too, of the part it plays in human culture.

No science educator, I should even more confidently add, can properly ignore the issues raised by philosophy of science. In any system of teacher training, 'philosophies-of' ought to play a central role. But that entails, if it is to be at all satisfactory, some introduction to *general* philosophy at the undergraduate level. Otherwise it is liable to collapse into superficial generalities, mere journalism or, even worse, mere preaching.

Australian National University

NOTES

[1] W. H. Watson (1963) *Understanding Physics Today*, Cambridge, pp. xi–xii.
[2] I have discussed such criticisms in my (1974) *Man's Responsibility for Nature*, London, and (1978) *Science and its Critics*, London.
[3] Hilary Putnam (1975) 'Philosophy of Physics', in *Mathematics, Matter and Method*, Cambridge, Vol. 1, p. 79.
[4] See, for example, Karl Popper (1972) *Objective Knowledge*, Oxford, p. 142.
[5] 'A Philosopher Looks at Quantum Mechanics', in *Mathematics, Matter and Method*, Vol. 1, p. 157.
[6] (1966) *Philosophical Foundations of Physics*, New York, p. 291.
[7] Mario Bunge (1968) 'The Maturation of Science', in *Problems in the Philosophy of Science*, ed. Imre Lakatos and Alan Musgrave, Amsterdam, p. 142.
[8] See especially Michael Ruse (1973) *The Philosophy of Biology*, London.
[9] Lawrence Sklar (1974) *Space, Time and Spacetime*, Berkeley, p. 417.
[10] One session of the 1978 Düsseldorf World Congress of Philosophy was devoted to the control of technological progress. Not a single Anglo-American Philosopher spoke in the course of the lengthy discussion which followed, although, admittedly, I chaired the proceedings and Rescher and Bunge read papers.
[11] (1973) *Man, Science and Technology*, produced by the Institute of Philosophy in the USSR Academy of Science, the Institute of Philosophy and Sociology in the Czechoslovak Academy of Sciences and the Institute of the History of Natural Sciences and Technology of the USSR Academy, Moscow/Prague, p. 313.
[12] For a fuller discussion of this confusion, compare John Passmore (1980) *The Philosophy of Teaching*, London.

[13] John Passmore, *Science and its Critics*, pp. 52–67.

[14] For a fuller discussion of Popper, his critics and his followers, see Alan Chalmers (1976) *What Is This Thing Called Science?* Brisbane, which is almost wholly devoted to epistemic philosophy of science.

[15] (1974) 'Reply to My Critics', in *The Philosophy of Karl Popper*, ed. P. A. Schilpp, La Salle, Illinois, Vol. 2, p. 1036. See also (1972) *Objective Knowledge*, Oxford, with its attempt to locate not only scientific conclusions but scientific problems in a 'third world'.

[16] Paul Feyerabend (1975) *Against Method*, London, p. 23.

[17] Herbert Feigl (1974) 'Empiricism at Bay? Revisions and a New Defense' in *Methodological and Historical Essays in the Natural and Social Sciences*, ed. Robert S. Cohen and Marx W. Wartofsky, Boston, p. 1.

[18] Ernest Nagel (1961) *The Structure of Science*, New York, p. 13.

[19] On the ground that it abrogates the critical function of philosophy. See Jürgen Habermas (1968) *Knowledge and Human Interests*, Frankfurt; Eng. trans. Boston, 1971. But one finds a not altogether dissimilar approach in Althusser's Marxism.

[20] Karl Popper (1963) *Conjectures and Refutations*, London, pp. 100–101. The opposite view is however taken by Paul Feyerabend in his 'On a Recent Critique of Complementarity', *Philosophy of Science*, 35 (1968), 309–311, 36 (1969), 82–105.

[21] J. G. Kemeny (1959) *A Philosopher Looks at Science*, Princeton, p. 31.

[22] N. R. Hanson (1962) *Patterns of Discovery*, London, p. 72.

EVERETT MENDELSOHN

KNOWLEDGE AND POWER IN THE SCIENCES

In the fall of 1947, J. Robert Oppenheimer, the man who directed the
U. S. project that made the atomic bomb, delivered the Arthur D. Little
Memorial Lecture at the Massachusetts Institute of Technology. In the
lecture, he said:

> Despite the vision and far-seeing wisdon of our war-time heads of State, the physicists
> felt a peculiarly intimate responsibility for suggesting, for supporting, and in the end
> in large measure for achieving the realization of atomic weapons. Nor can we forget
> that these weapons, as they were in fact used, dramatized so mercilessly the inhumanity
> and evil of modern war. In some sort of crude sense which no vulgarity, no humor, no
> over-statement can quite extinguish, the physicists have known sin, and this is a knowl-
> edge which they cannot lose.[1]

Physicists have known sin. Sin came to them, in Oppenheimer's view, through
power, the power that science had gained and used.

   Going back over 300 years to the beginning of the period which we refer
to as the Scientific Revolution, we find that the image of power was put very
clearly and very directly by Francis Bacon, philosopher, Lord Chancellor, a
commentator on knowledge and what it might do. Let me quote him.

> It is well to observe the force and effect and consequences of discoveries. These are to be
> seen nowhere more conspicuously than in those three which were unknown to the
> ancients and of which the origin, though recent, is obscure, namely, printing, gun
> powder, and the magnet. For these three have changed the world: the first in literature,
> the second in warfare, the third in navigation, whence have followed innumberable
> changes. In so much that no empire, no sect, no star seems to have exerted greater
> power and influence in human affairs than these mechanical inventions.[2]

"Knowledge is power," Bacon said in his famous epigram. "We understand
nature in order to command her" is another well worn Baconian phrase. The
goal was human gain, to improve the commonweal, but the basic aim was to
have dominion over things. The new science, the new knowledge was, indeed,
to improve the estate of humans on earth. But there was something more
behind Bacon's science, and, in a sense, it had a power over him, as it was to
have over his contemporaries. In developing this new approach to nature,
these new ways of controlling nature, humans, he felt, would regain their
prelapsarian state of the period before the fall.[3] They could prolong life,

31

R. W. Home (ed.), Science Under Scrutiny, 31–47.
© 1983 by D. Reidel Publishing Company.

they could conquer disease, they could do all those things which humans had dreamed of, but which they had not been able heretofore to achieve.

More than 300 years separate Bacon and Oppenheimer. The earlier vision of Bacon was an optimistic one but, curiously enough, it was developed in a time when humans actually possessed very little power to control nature. The more recent vision is a pessimistic one. At the very time that humans have achieved the ability to exert power over nature, they question the ends of such ability. What has happened in between? Why has this change of outlook and change of attitude occurred?

There are two clear senses in which the notion of power is used. One reflects science's ability to do powerful things, to bend nature to the human will. The second refers to the achievements of scientists and to their positions of power and authority in the societies of which they are a part. Whence does science derive its power? How has it developed?

There are at least four ways of looking at science and its sources of power. First, science is a body of concepts and techniques involving theories and ways of acting and thinking. Second, science is a way of knowing, a way of ordering reality and, ultimately, of acting upon that reality. It provides a way of knowing that has transcended the boundaries of the practising scientist and has been adopted broadly in society at large. Third, science is a socially organized activity, a profession with a locus within society and part of a hierarchy, related to other social institutions including governmental, military, and industrial. Finally, science is a source of utilitarian instruments of power for science and for others to use.

The seventeenth century was a period of fertile, conceptual innovation. Indeed, the innovations in ways of knowing and in conceptual structures of this period are what we label as the Scientific Revolution. An epistemology was established which set in place a way of knowing nature through reason and experience. Processes of demarcation were established to set rules about what was to be included in the sciences and what was to be excluded from them. Both the cognitive nature and the content of science were being developed.

Rules of exclusion from science were important. There were certain topics science would eschew. Science would not deal, said the founders of the Royal Society of London (and indeed, this might have been repeated by the founders of almost any of the academies of science in the seventeenth century) with religion, rhetoric, metaphysics, politics, or morality.[4] But even though it claimed it would not deal with these topics, science did maintain one value quite clearly — dominion over nature or mastery over things. While moral or normative guides were eschewed, the notion of mastery was central. Historians have called this the positivist compromise.[5]

The seventeenth century was also the period when the social differentiation of the scientists' role was begun.[6] Science was institutionalized and its organizations established.[7] It became separated from other activities within society. Scientists could be recognized as a group called natural philosophers or experimental philosophers and clearly separate from theologians, princes, and merchants. The institutionalization also provided a putative means of resolving conflict and of establishing truth. New scientific societies and academies, as they were established in the seventeenth century, provided the institutional bases within which new methods were developed. It was in the context of these societies that debates about nature were to be found. The role was developed of scientific peers as challengers of findings, and the need to provide reproducible results was recognized. These societies were responsible for the introduction of processes of control — control which was both social and cognitive — over the ways in which scientists acted and thought. They established the content, boundaries, and procedures of science. Individuals gave over their authority to these societies at some level while the societies, in turn, developed interests of their own. These institutional interests became reflected in the way scientists acted and in the way they responded to other members of society.

This was the period when ties to the state for support and for financial patronage were formed. In Italy, France, Germany and Russia, the state — or its agents, the aristocracy — were the direct patrons and controllers of science during much of the seventeenth and eighteenth centuries.[8] Only in England was the situation different. There, while scientists gained a Royal Charter in their early years, they gained nothing in the way of royal patronage or actual support.[9] Nevertheless, we can claim that even in England, this is the time when science became recognized and empowered officially and informally.

But what about power — Bacon's goal? During the seventeenth century, science's ability either to command nature or to gain power of a social sort was extremely limited. Few were in science's debt, though science was in debt to many others. The interests of science, at this time, are largely those of an esoteric or organizational kind. Society had some specific interests in what science could do, for example in astronomy and its aid to navigation. But, by and large, the social interests surrounding science were narrower ones, of the group of people organized together to carry out the activity of science. They were largely from the educated upper-middle classes, and they sought to achieve space and recognition within their society through their work.[10]

Science had proclaimed its utility to society, and its practitioners attempted to demonstrate this utility in order to enhance their roles. Furthermore, science, at the time, largely avoided conflict with the established power of the state and the church. Bishop Thomas Sprat's history of the Royal Society, written a few years after its founding in the 1660s, is largely a treatise of apologetics supporting the promise that science would in no way tread on the toes of the state or the church.[11] We can conclude, therefore, that the cognitive features of science yielded little actual power. No sovereigns feared it, and several were involved in taming it. The church, of course, subdued it at its will, as the Counter Reformation made abundantly clear.[12]

The eighteenth century witnessed a further transformation in the scientific way of knowing, a transformation particularly involving the public. This change was largely outside the boundaries of science itself. Both the new politics of the Age of Reason and the new politics of the nascent industrial revolution gave legitimacy to the scientific way of knowing and to the belief that one could act on nature in an effective and economically viable manner. Scientists themselves were only partially cognizant of the new way of looking at nature that was taking shape outside their ranks as part of the growth in the industrial societies. The sources of industrially oriented experimentation have been examined by Henry Guerlac in his studies of the development of industrial chemistry in France during the eighteenth century.[13] The people involved were at one and the same time practising chemists and developers of the theoretical bases of what emerged by the end of the century as the Chemical Revolution. This same approach appears in the works of Condorcet, the great philosopher of the French enlightenment, who carried philosophical and educational visions into the revolution itself.[14] Science represented, for him, the highest point of progress of the human race.

The provincial manufacturers in England also responded to these new ways of dealing with nature. Textiles were their focus, and dyes and mordants were the chemical agents they needed. For leather working and for metals, they needed acids and alkalis. As a means of furthering their interest in studying those sciences that were allied to the new industries, they set up new organizations in the provincial centers — Manchester, Birmingham, Newcastle, Leeds.[15] Through these new societies, not only did they gain their practical view of what the sciences had to offer but, as Arnold Thackray has shown of the Manchester Literary and Philosophical Society, they also began using science as the basis of a new culture for the industrial middle classes.[16] They were, after all, outside the bounds of the upper classes of British society and, as such, divorced from belles-lettres. For them, science provided the basis of a new culture. In all of these cases it was the uses of science that were most clearly visible and thus most widely known among the public.

The dissemination of science as a way of knowing was also spread among the public at large. Humphry Davy caught this new mood at the turn of the century, in planning improvements to the Royal Institution in London. He hoped that the practical worker could benefit from being instructed as to the correct scientific theories of his particular branch of labour and that he, in turn, would freely communicate his methods to the philosophical inquirer so that they might be corrected by scientific principles.[17] This represents a clear attempt to make science directly relevant to industrial pursuits. From this diffusion of knowledge and from its utilization, whatever power science exercised during the course of the late eighteenth and early nineteenth centuries was derived. This dissemination of a way of knowing represented a form of politicization and opened the way for potential external determination of the shape and actions of science.

During the course of the nineteenth century we witness what may be characterized as struggles for turf. This was the period of the establishment of the professional characteristics of science. It was also a period of re-establishing cognitive boundaries. During the nineteenth century, particularly during its early decades, every one of the specific disciplines, from anatomy and astronomy to statistics and zoology, established disciplinary identity. The A to Z of the sciences was set out in a cognitive framework with the new disciplines defining the problem areas within which scientists would work. In addition, each discipline formed its own professional societies and all the disciplines campaigned for the establishment of new specialized university professorships and institutes.[18]

This was also a period of working out the social boundaries of science. Who would be included and who excluded? Closely connected with the move to professionalism, there was a move to reset science in a social locus within society that would give it position, status and, ultimately, power. The heightened pace of the Industrial Revolution enhanced the recognition of the utility of scientific knowledge. From the turn of the century it became clear that science would play a basic role in the new industries. This was the message that emerged from the founding of the École Polytechnique in Paris in the closing years of the French Revolution, where science was linked directly to technical, industrial and military pursuits. Similarly it was the message that in England motivated the establishment of the Royal College of Chemistry and the Royal School of Mines in the early decades of the nineteenth century. Both these institutions were directed toward the practical applications of science and both tied education in the sciences to industrial development. The scientists of this generation stressed the role they could play. In 1853, for example, George Hopkins, in his presidential address to the recently founded British Association for the Advancement of Science, put the utilitarian claim very directly: "One great duty we owe to the public is to encourage the application of science to the practical purposes of life, to bring, as it were, the study and the laboratory into juxtaposition with the work shop."[19] The image thus created was of the laboratory and the work shop side by side.

But even as scientists provided industrial power and gained new stature in so doing, they attempted to obscure the very interests which were responsible for their position and status. It is at this point in history that we see the beginning of an internal tension which has remained with science to the present day. Scientists attempted to back away from the implications of the utility which had now become obvious and which was often influential in the cognitive processes in which they were engaged. Those interests became more noticeable in the prolonged and intense debates and controversies which took place in science during the nineteenth century. Richard Owen, for example, recognized what lay beneath the surface in the French debate between Pasteur and Pouchet and the earlier debate between Cuvier and Lamarck and Geoffrey St. Hilaire. He put the matter very bluntly: "Pasteur, like Cuvier, had the advantage of subserving the prepossessions of the party of order and the needs of theology."[20] Power was being gained through serving established authority in church and state. Science's position was improving, its power increasing and its ties to the state becoming clearer.

During this period of explicit attempts to utilize science in new industries such as synthetic organic chemicals and electricity, there was also an explicit move to identify 'pure' science as opposed to 'applied' science. The very call on the word 'pure' and the invocation of purity suggests by implication, of course, that the non-pure is somehow contaminated. There was certainly an attempt to give some special sanction and special legitimacy to the newly defined 'pure' science. Prince Albert, one of the great patrons of science in the mid-century in Britain, set the tone for describing the 'pure' scientists' task. He said they should accept "a self-conscious abnegation for the purpose of portecting the purity and the simplicity of their sacred task."[21] The words are clear — science was something very special, something very strong, something separated from other activities. Lyon Playfair, an industrial chemist who himself had gone to Parliament and then the University, put the issue equally clearly. What is it that science should look like? Who should be rewarded? "The discoverer of abstract laws," he wrote, "is the real benefit of his kind — far more than he who applies them directly to industry."[22] The aim of science? Finding new knowledge, he said. The true scientists, he said, were men who are "looking for sublime truths, careless of whether they will have any immediate effect on industry." It is just this vision of the separation of the utility from the knowledge itself which has since come under fire from Jürgen Habermas: "Because science must secure the objectivity of its statements against the pressures and seduction of particular interests, it deludes itself about the fundamental interest to which it owes not only its impetus, but the conditions of possible objectivity themselves."[23] It is curious, then, that even as science became industrially vital, an important part of the productive forces of society in the 19th century, it attempted to 'denormatize' itself, to take away from itself any vestige of interest in utility or in other values which might guide it. In a sense, science seemed to want to cleanse itself of interests, even of narrow professional interests.

William Whewell, the mathematician-philosopher, put it very bluntly. "Knowledge, is power", he wrote in mid-century. "But," he added, "for us it is to be dealt with as the power of interpreting nature and using her forces, not as the power of exciting feeling of mankind and providing remedies for social evils on matters where the wisest men have doubted and differed."[24] This is the separation. Knowledge was to be pulled away from the very power which it was gaining. Yet power itself, and the interests involved with it, deeply influenced the cognitive patterns and choices that scientists were making. Both professional interests and broader social interests — of theology, of the state, and of industry — were guiding science, shaping it, and molding it in these years. Its place in society; the things it could do; the tasks it would undertake; the vision of what it should do for its students and its practitioners; all were being shaped by the new power which it was gaining. But, even as science gained power, the process was not unidirectional. The interests which affected it were not an intrusion into an otherwise pure and rational procedure. Even while avoiding sociological reductionism, we must recognize the tension that came to exist as knowledge gained and used power — the tension that came to exist between science on the one hand and society on the other, between the cognitive and the social realms of scientific activity.

To the historian, it is curious that 19th-century science established its base of power in the ivy-covered towers of universities. From the 17th century, science had really not been an important force in the universities, and most science was carried on outside their walls during these two formative centuries. With great pressure, science managed to enter the universities in the 19th century and, as we know, its role in the universities was greatly resented by the traditional university inhabitants. One of these, the mathematician Charles Dodgson (better known as Lewis Carroll), wrote a brilliant and critical letter to the *Pall Mall Gazette* in 1872 in which he showed the kind of resentment that the traditional member of the university had toward science. For its wit and wisdom I quote from this:

Let me sketch in dramatic fashion the history of science's recent career at Oxford. In the dark ages of our University, some five and twenty years ago, while we still believed in classics and mathematics as constituting a liberal education, natural science sat weeping at our gates. 'Ah, let me in,' she moaned, 'Why cram reluctant youth with your unsatisfying law? Are they not hungering for bones, yea panting for sulphurated hydrogen?' We heard and we pitied. We let her in and we housed her royally. We adorned her palace with reagents and retorts and we made it a very charnel house of bones. And we cried out to our under-graduates, 'The feast of science is spread. Eat, drink and be happy.' But they would not. They fingered the bones and they thought them dry. They sniffed at the hydrogen and turned away. Yet for all that science ceased not to cry, more gold, more gold. And her three fair daughters — chemistry, biology and physics, for the modern horse leach is more prolific than in the days of Solomon — cease not to plead, give, give. And we gave. We poured forth our wealth like water. I beg your pardon, like H_2O. And we could not help thinking there was something weird and uncanny in the ghoul-like facility with which she absorbed it.[25]

But, in spite of Dodgson's opposition and that of other traditionalists, science did enter. We can see an almost dialectical process occurring. Even as science gained in utility to industry, it took flight into the universities. The tension between professional interests on the one hand and industrial interests on the other grew, even though these interests were never totally separated.[26] This can be seen in the transformations that occurred in institutional forms, with the establishment of *technische Hochschulen* in the German-speaking world, technical institutes in the English-speaking world, polytechnics in the French world. But the interests were set in a hierarchy with pure scientists on top and located within the universities, not in the institutes.

To understand science in the universities, we can take an archaeological approach. In your mind, wander around one of the older universities. The newer buildings are the ones that scientists inhabit. In older universities, these form a ring around the central core. They are larger than the buildings of other faculties. New curricula were developed. The sciences developed a research emphasis, replacing the moral training or the tradition-building that had been the role of the universities. New institutions were formed, often on the periphery of the universities, that later moved into and reformed the universities themselves.

The 19th century was also a period of professionalization.[27] The pattern of professionalized science was to drive out the amateur and to declare roles and ways in which scientists should act. The pattern spread through Europe to the United States and to all parts of the scientific world, and it enormously strengthened science organizationally. Having entered the universities, science gained the ability to train the new generation, to set the norms and the problems and the patterns that the new generation should act by. This professionalization is important to the process of making organized science a powerful social institution as well as a producer of power for other institutions and societies.

But remember the words of George Bernard Shaw: "The professions," he proclaimed, "are a grand conspiracy against the laity." Institutional and professional power were gained at social cost. What scientists applauded — the increased differentiation, the new status, the reduced amateurism — were seen as a threat by lay segments of the public. To have power is to become suspect. So as science gained its power, it also gained its critics. The vivisection controversy that developed in the closing decades of the 19th century provides an example. The clear professional interest was to do experiments with living animals as a means of bettering the position of the physiologists and of advancing their knowledge. Experiments with live animals, however, came directly into confrontation with broadly held public mores having to do with the humane treatment of animals. Also, interestingly, women were strong among the antivivisectionists. One of the leaders, Frances Power Cobb, was not only an antivivisectionist but also a vigorous suffragette and a feminist. She and others like her who joined the antivivisection movement did so because of their distrust of physicians who were always male and who, they felt, did not understand women's diseases. The antivivisection movement also gained adherence from others who were opposed to the institutionalised power of science as they saw it in things such as compulsory vaccination. They saw a scientific establishment, and they wished to challenge it.[28]

The 19th century sees the invention of whole new departments in universities and whole new Institutes with ever higher thresholds for entry. A pattern of exclusivity and inaccessibility thus developed in which special knowledge is gained and used by special people

At just this time, however, there is also the movement toward popularization in the sciences. Thomas Henry Huxley lectures to the working man. John Tyndall in England, Emil Du Bois-Reymond in Germany, and numerous others attempted to bring science to the public.[29] These efforts reflect an ambivalence on the part of scientists. Should the separation of scientific knowledge really occur, and, if it did, what would the consequences be? How could one avoid it? Mechanics' institutes were established which aimed to bring knowledge of the sciences to those who needed technical training.[30] With them came a socialization to technical norms through which skilled workers accepted the patterns being developed by the experts. Other public or semi-public institutions that aimed in some way at bringing the sciences to different groups were also developed. The Royal Institution, for example, was established at the beginning of the century. Its exhibits and public lectures were initially aimed at mechanics and skilled artisans, but it ended up providing lectures to white-tie audiences drawn from London's upper classes and social élite.[31] In France, the famous Conservatoire des Arts et Métiers in Paris and numerous copies of it in other French centres were established.[32] However, while the technically trained scientists and engineers thus brought their material, their views and their outlook to the public, the public had no way of speaking directly to the scientists or the experts. The feedback loop was not closed.

Why do I dwell on these aspects of separation and antagonism? Because the slow creation of a knowledge élite, as science became more complex and harder, also resulted in the creation of a social élite. Though scientists were not yet found at the highest points of social power, they had become, as C. Wright Mills, the American sociologist, called them, "the technical lieutenants of power," the people just below the top level who would exercise power in concert with the leaders of the state.

The 20th century is an epoch of experts and their powers — and their problems. By the end of the 19th century, modern states had recognized the need for the sciences. Prince Albert, in his presidential address to the British Association in 1859, foresaw the day when science "will no longer require the begging box, but will speak to the state like a favoured child, sure of his paternal solicitude for his welfare."[33] Indeed, the position of science changed and, though the patronage of the state came slowly, it did come.

The earliest state patronage, and the earliest influence, had come through new industries, especially those which gave basic things such as the new source of power in electricity.[34] The electricity industry was dependent upon and closely linked to the sciences and the public mind knew of this link. Though scientists preferred to ignore the closeness of this tie, popular literature, popular illustrations and popular books of the day showed it over and over again. The public saw science as giving it something it wanted and providing national strength. By the year 1900 all things were felt to be knowable and all problems solvable if only science could be put to work. The interests of the scientists became identified with industrial success. Scientism became the outlook of the modern industrial societies.

This form of legitimation — that is, giving sanction to science because of its productive abilities — worried some of the scientists themselves. They were afraid of the ultimate feedback. With industrial failure such as occurred during the great Depression of the 1930s, science could be seen as being at fault. During the 1930s, in Britain, in the United States, and on the Continent, calls were heard for moratoria on the gaining of new knowledge and technique.[35] Stop at this new knowledge; stop at this new technique. Let us learn to use what we have; let us set our house in order before we seek after further innovations. In a peculiar way this attitude was mirrored within science. Some scientists called for the transformation of the practice of science to aid the creation of a new social order — the social order of socialism. Among a group of British scientists in the 1930s, a movement arose to use science in a socially progressive manner.[36] One critic suggested that the House of Lords should be abolished in favor of a senate of scientists to deal with the real problems in society. Here we see normative views merging explicitly with social images and social values compelling scientists to redefine what it was that they felt science could and should do.[37] This debate during the 1930s can be seen against the background of the idealized image of science which the protagonists saw as being at the core of the state in the Soviet Union. The implications were great when the Soviet Union became a focus of criticism later.[38]

In the 1930s, then, we see the beginning of a process of de-institutionalization and de-professionalization of science. Consensus within the scientific community and the organized patterns of science itself began to break down. Scientists and others were willing to step outside the professional boundaries and become critics of the material of science and the activities in which it was engaged. The attitudes that developed toward science during the years of the Depression, both within the scientific community and without, were a harbinger of what was to come. The major turning point, however, came in the decision to make and use the atomic bomb. Samuel K. Allison, one of the American physicists directly involved in the bomb project, drew the dire conclusion that physics and physicists were important to the state for waging war. Edwin U. Condon, another of the key figures in that activity, wondered whether a revulsion toward war that would follow the use of the bomb might extend to the sciences.[39] Indeed, it was exactly this revulsion to war, and the linking of the sciences to war, that emerged more strongly in the Vietnam decade and affected the attitudes of many young people toward science and scientists in the universities.

Science, through the making of the bomb, had become politically interesting. And, because of their involvement in making the bomb, scientists themselves became interested in politics. The whole culture of science as it developed in the post-war years was subject to question. People asked, "What about the scientists who engaged in the actual making of the bomb, in the days just before its use? Were there no critics?" There were some, we know. But, as Oppenheimer testified, they went ahead with making the bomb and not asking the moral questions about its potential use because the problem was "technically sweet." It is just this view that Freeman Dyson, a mathematical physicist at the Institute for Advanced Study at Princeton, has captured in a recent article in *The New Yorker*. "Nuclear explosions," he writes, "have a glitter more seductive than gold to those who play with them. To command nature, to release in a pinpoint the energy that fuels the stars, to lift by pure thought a million tons of rock into the sky — those are exercises of the human will which produce an illusion of illimitable power."[40] Dyson looks at the roles of Oppenheimer and Edward Teller. Though recognizing the contrast and conflict between these two, he writes,

I believe that each of them, having achieved his technical objective, wanted more. Each of them sought political as well as technical power; each of them became convinced that he must have political power to ensure the direction of the enterprise he had created should not fall into hands that he considered less responsible. In the end, each of them was irrevocably committed to the exercises of the human will in the political as well as the technical sphere.

Each with the "illusion of illimitable power" sought, also, political power. What is it that power does?

At the same time, other factors were changing the nature of the relationship between the expert and the lay person. The very successes in the education of the public in the sciences created the potential for their demystification. The public gained the potential for being less afraid of questioning the experts.

Deep political trends also led to a questioning of the experts. The 'politics of participation' swept Europe and the United States — in the U.S. in the Civil Rights Movement and the Vietnam War protests, on the Continent in the breakdown of the authoritative relations that had been in place since the end of the Second World War. Science was challenged, both as a source of power and as a social élite. Its candor and its honesty were questioned; its interests and its ties were explored. One of the men who set the clearest guide for this critical exploration was Dwight D. Eisenhower. Aware of the locus of power, he gave focus to the relationship between industry, scientific experts, and the military, and he warned against what might come of their continued alliance.[41]

During the 1960s and the early '70s, the challenge to science went further. Its critics challenged the very epistemology, the very way of knowing of science itself, especially science's inability and failure to deal with normative or value issues.[42] The claim of being value-neutral or value-free was scoffed at as science was seen to use its power in negative ways.[43] Which interests, whose interests, were in command? Practitioners of science themselves, and science watchers — historians, philosophers, sociologists, and political scientists — joined in the challenge. Our discussions have represented an attempt to understand and reconstruct the very activity that we call science. All the major issues are open. The critical questions are being asked and authority is no longer in place. There is danger in too quickly forcing closure of these debates, for if there is one thing to be learned from examining the climb of science from its will to power to its exercise of power, it is that the enterprise of science is too powerful to leave to the experts. The role of the citizen needs full exploration.

Harvard University

NOTES

[1] J. Robert Oppenheimer (1947) 'Physics in the Contemporary World', reprinted in *The Open Mind*, New York, 1955.

[2] Francis Bacon, *The New Organon*, Bk. I, Aphorism CXXIX.

[3] See Frances A. Yates (1967) 'The Hermetic Tradition in Renaissance Science', in Charles S. Singleton (ed.), *Art, Science, and History in the Renaissance*, Baltimore, pp. 266–267; see also Paolo Rossi, (1957) *Francis Bacon: From Magic to Science*, trans. by S. Rabinovitch, London, 1968; see Bacon's 'Preface' to *The Great Instauration* for a discussion of the relation of knowledge to The Fall and the terms he recommends for beneficial human acquisition of knowledge.

[4] See Robert Hooke (1848) 'The Business and Design of the Royal Society', in R. Weld, *A History of the Royal Society, With Memoirs of the Presidents, Compiled from Authentic Documents*, London, 2 vols. Cf. the 'Preface to the Reader' to Accademia del Cimento, *Essayes of Natural Experiments* (1667) trans. by Richard Waller, London, 1684.

[5] For a summary of this discussion see Everett Mendelsohn (1977) 'The Social Con-

struction of Scientific Knowledge', in E. Mendelsohn, P. Weingart and R. Whitley (eds.), *The Social Production of Scientific Knowledge, Sociology of the Sciences Yearbook*, Vol. 1, pp. 3–26. The concern with *Power* and control as themes in relation to knowledge has emerged as a particular focus in the recent works of Michel Foucault; see the recent collection of essays edited by Colin Gordon, (1980) *Michel Foucault, Power/ Knowledge: Selected Interviews and Other Writings, 1972–1977*, New York. *Domination, Dominion* and *Mastery* over nature have been treated in several recent studies; see Clarence J. Glacken (1967) *Traces on the Rhodian Shore*, Berkeley, especially Chapter 10, 'Growing Consciousness of the Control of Nature'; William Leiss (1972) *The Domination of Nature*, New York; Charles Webster (1975) *The Great Instauration: Science, Medicine and Reform, 1626–1660*, London, especially Chapter 5 'Dominion Over Nature'; Carolyn Merchant (1980) *The Death of Nature*, New York, especially Chapter 7, 'Dominion Over Nature'.

6 Joseph Ben-David (1971) *The Scientist's Role in Society, A Comparative Study*, Englewood Cliffs, N.J., especially Chapters 4 and 5.

7 The fullest examination remains that of Martha Ornstein (1913) *The Role of Scientific Societies in the Seventeenth Century*, New York; rev. ed. Chicago, 1938. A thoughtful and detailed social history of the Paris Academy is Roger Hahn (1971) *The Anatomy of a Scientific Institution: The Paris Academy of Sciences, 1666–1803*, Berkeley. A still valuable history of fledgling efforts in France is Harcourt Brown (1934) *Scientific Organizations in Seventeenth Century France (1620–1680)*, New York; rev. ed. 1967. For a view of what happened outside the major centres of learning, see K. Theodore Hoppen (1970) *The Common Scientist in the Seventeenth Century: A Study of the Dublin Philosophical Society, 1683–1708*, Charlottesville, Virginia. The scientific setting in England is thoughtfully examined in Michael Hunter (1981) *Science and Society in Restoration England*, Cambridge.

8 See Martha Ornstein, *Role of Scientific Societies*, for a comparative assessment; for other comparative materials see Maurice Crosland (ed), *The Emergence of Science in Western Europe*, New York, 1976; London, 1975.

9 Henry Lyons (1944) *The Royal Society, 1660–1940: A History of Its Administration Under Its Charters*, Cambridge; Margery Purver (1967) *The Royal Society: Concept and Creation*, Cambridge, Mass.

10 For social class origins of scientists see Robert K. Merton (1938) 'Science, Techology and Society in Seventeenth Century England', *Osiris*, 4, Pt. 2; and Nicholas Hans (1951) *New Trends in Education in the Eighteenth Century*, London.

11 Thomas Sprat (1959) *History of the Royal Society*, edited with critical apparatus by Jackson I. Cope and Harold Whitmore Jones, St. Louis.

12 Giorgio de Santillana (1955) *The Crime of Galileo*, Chicago; Jerome J. Langford (1971) *Galileo, Science and the Church*, Ann Arbor, rev. ed. For a sympathetic explanation of the attitudes of the Counter-Reformation, see François Russo (1963) 'Catholicism, Protestantism, and the Development of Science in the Sixteenth and Seventeenth Centuries', in Guy S. Métraux and François Crouzet (eds.) *The Evolution of Science*, New York, pp. 291–320.

13 Henry Guerlac (1959) 'Some French Antecedents of the Chemical Revolution', *Chymia 5*, pp. 73–112; see also A. E. Musson and E. Robinson (1969) *Science and Technology in the Industrial Revolution*, Manchester; David Landes (1969) *The Unbound Prometheus*, Cambridge; A. and N. L. Clow (1952) *The Chemical Revolution*, London;

Peter Mathias (1972) 'Who Unbound Prometheus? Science and Technical Change, 1600–1800', in Peter Mathias (ed.) (1972), *Science and Society, 1600–1900*, Cambridge.

[14] Keith M. Baker (1975) *Condorcet: From Natural Philosophy to Social Mathematics*, Chicago.

[15] See for example, Robert E. Schofield (1963) *The Lunar Society of Birmingham: A Social History of Provincial Science and Industry in Eighteenth-Century England*, Oxford; for an overview of trends see D. S. L. Cardwell (1957) *The Organisation of Science in England: A Retrospect*, London; Robert H. Kargon (1977) *Science in Victorian Manchester: Enterprise and Expertise*, Baltimore; for a comparative view of the French scene see Maurice Crosland (1967) *The Society of Arcueil: A View of French Science at the Time of Napoleon I*, Cambridge, Mass.; for the United State see Alexandra Oleson and Sanborn C. Brown (eds.) (1976) *The Pursuit of Knowledge in the Early American Republic: American Scientific and Learned Societies from Colonial Times to the Civil War*, Baltimore.

[16] Arnold Thackray (1974) 'Natural Knowledge in Cultural Context: The Manchester Model', *Amer. Hist. Rev.* 79, pp. 672–709.

[17] Humphry Davy (1810) *A Plan for Improving the Royal Institution and Making it Permanent*, a lecture delivered 1810, London, p. 10; Morris Berman (1978) *Social Change and Scientific Organization: The Royal Institution 1799–1844*, Ithaca, N.Y.

[18] For an examination of the moves toward 'professionalization' in the sciences see Everett Mendelsohn (1964) 'The Emergence of Science as a Profession in Nineteenth-Century Europe', in Karl Hill (ed.), *The Management of Scientists*, Boston, pp. 3–48; see also the critical discussion in Joseph Ben-David (1972) 'The Profession of Science and Its Powers', *Minerva* 10, pp. 362–383.

[19] George Hopkins (1853) *Proceedings, BAAS*, p. lvii; the role of the BAAS has been the focus of an important new study, Jack Morrell and Arnold Thackray (1981) *Gentlemen of Science: Early Years of the British Association for the Advancement of Science*, Oxford.

[20] John Farley and Gerald Geison (1974) 'Science, Politics and Spontaneous Generation in Nineteenth-Century France. The Pasteur-Pouchet Debate', *Bull. Hist Med.* 48, p. 167.

[21] Albert, Prince Consort (1862) 'Lecture Delivered at the International Statistical Congress, July, 1860', *The Principal Speeches and Addresses of the Prince Consort*, London, p. 255.

[22] Lyon Playfair (1852) *Of the Chemical Principles Involved in the Manufactures of the Great Exhibition as Indicating the Necessity of Industrial Expansion*, London, p. 9.

[23] Jürgen Habermas (1971) *Knowledge and Human Interests*, Boston, pp. 301–317.

[24] William Whewell (1841) *Proceedings, BAAS*, p. xxxiv.

[25] Charles Dodgson, excerpts from 'Natural Science at Oxford', Letter to the Editor of the *Pall Mall Gazette*, May 12, 1872.

[26] Education in and for the sciences and engineering has recently become the focus of active scholarship. See John Hubbel Weiss (1982) *The Making of Technological Man: The Social Origins of French Engineering Education*, Cambridge, Mass.; Terry Shinn (1980) *L'École Polytechnique 1794–1914: Savoir Scientifique et Pouvoir Social*, Paris; Gordon Roderick and Michael D. Stephens (1972) *Scientific and Technical Education in Nineteenth-Century England*, Newton Abbot; Michael Sanderson (1972) *The Universities and British Industry 1850–1970*; Fritz K. Ringer (1969) *The Decline of the German Mandarins*, Cambridge, Mass.; see also the fine series of studies edited by Robert

46   EVERETT MENDELSOHN

Fox and George Weisz (1980) *The Organization of Science and Technology in France, 1808–1914*, New York.

[27] See Note 18 above.

[28] See some of my remarks on this controversy in Everett Mendelsohn (1983, in press) 'The Political Anatomy of Controversy in the Sciences', in Arthur L. Caplan and H. Tristram Engelhardt, Jr. (eds.), *Scientific Controversies: Studies in the Resolution and Closure of Disputes Concerning Science and Technology*, New York; for a fine detailed history see Richard D. French (1975) *Antivivisection and Medical Science*, Princeton.

[29] Popularisation of the sciences is a field crying out for serious social-historical examination. One of the few analytical studies is Philippe Roqueplo (1974) *Le Partage du Savoir: Science, Culture, Vulgarisation*, Paris.

[30] For a thoughtful analysis see Steven Shapin and Barry Barnes (1977) 'Science, Nature and Control: Interpreting Mechanics' Institutes', *Soc. Stud. Sci.* 7, pp. 31–74.

[31] Morris Berman, *Social Change and Scientific Organization*, see Note 18 above.

[32] Robert Fox (1974) 'Education for a New Age: The Conservatoire des Arts et Métiers, 1815–1830', in D. S. L. Cardwell (ed.), *Artisan to Graduate*, Manchester.

[33] Albert, Prince Consort (1859) *Proceedings, BAAS*, p. lxviii.

[34] A good introduction to the issue of the support of science in England is Roy M. MacLeod (1972) 'Resources of Science in Victorian England: The Endowment of Science Movement, 1868–1900', in Peter Mathias (ed.), *Science and Society, 1600–1900*, Cambridge, pp. 111–166; see also Gerard l'E. Turner (ed.) (1976) *The Patronage of Science in the Nineteenth Century*, Leiden.

[35] See Carroll Purcell (1974) ' "A Savage Struck by Lightning": The Idea of a Research Moratorium, 1927–37', *Lex et Scientia* 10, pp. 146–161.

[36] For a full study of this movement see Gary Werskey (1978) *The Visible College: The Collective Biography of British Scientific Socialists of the 1930s*, New York.

[37] I examine this theme in Everett Mendelsohn (forthcoming) 'Social Chaos and Scientific Utopia', in Everett Mendelsohn and Helga Nowotny (eds.), *Utopian Science and Scientific Utopias, Sociology of the Sciences*, Vol. 7.

[38] The issue was most sharply raised by the chemist Michael Polanyi (1940) *The Contempt of Freedom: the Russian Experiment and After*, London; and John R. Baker (1945) *Science and the Planned State*, London. The debate is studied in William McGucken (1978) 'On Freedom and Planning in Science: The Society for Freedom in Science, 1940–46', *Minerva* 16, pp. 42–72.

[39] See Daniel J. Kevles (1977) *The Physicists: The History of a Scientific Community in Modern America*, New York, pp. 369–370.

[40] *New Yorker*, 20 August, 1979, p. 41. This material was subsequently reused in his book, Freeman Dyson (1980) *Disturbing the Universe*, New York.

[41] Eisenhower used two phrases, one of which has been widely adopted – "military-industrial complex"; the second has been more cautiously received – "scientific-technological élite."

[42] The criticism of science developed as part of a broader critique of industrial society and its cultures and values. The best known of the 'public' critics is Theodore Roszak (1969) *The Making of a Counter Culture: Reflections on the Technocratic Society and Its Youthful Opposition*, New York. An analysis of the new critics is found in Helga

Nowotny and Hilary Rose (eds.) (1979) *Counter-Movements in the Sciences, Sociology of the Sciences*, Vol. 3.

[43] See Brian Easlea (1973) *Liberation and the Aims of Science: An Essay on Obstacles to the Building of a Beautiful World,* London; Rita Arditti, Pat Brennan and Steve Cavrak eds.) (1980) *Science and Liberation*, Boston; Jean-Mare Lévy-Leblond and Alain Joubert (eds.) (1975) *(Auto)critique de la Science*, Paris.

ALAN MUSGRAVE

FACTS AND VALUES IN SCIENCE STUDIES

In every *recent study of science*, which I have
hitherto met with, I have always remark'd, that
the author proceeds for some time in the ordinary way of reasoning, and establishes the
existence of scientific communities, or makes
observations concerning *their* affairs; when of
a sudden I am surpriz'd to find, that instead of
the usual copulations of propositions, *is*, and
is not, I meet no proposition that is not connected with an *ought*, or an *ought not*. This
change is imperceptible; but is, however, of
the last consequence. For as this *ought*, or
ought not, expresses some new relation or
affirmation, 'tis necessary that it shou'd be
observ'd and explain'd; and at the same time
that a reason should be given, for what seems
altogether inconceivable, how this new relation
can be a deduction from others, which are
entirely different from it. But as authors do not
commonly use this precaution, I shall presume
to recommend it to the readers; and am per-
suaded, that this small attention wou'd subvert
all the vulgar systems of *philosophy and history
of science*, and let us see, that the distinction
of *scientific* vice and virtue is not founded
merely on the *social* relations of *scientists* ...
(DAVID HUME, *A Treatise of Human Nature*
(1739), III, i, 1; my italics).

1. THE PROBLEM OF SCIENTIFIC PROGRESS

'What's so great about science?', asks Paul Feyerabend, and does not stay long
for an answer. Convinced that none can be given, he bemoans the fact that:

... while an American can now choose the religion he likes, he is still not permitted to
demand that his children learn magic rather than science at school. There is a separation
between state and church, there is no separation between state and science. ... Physics,
astronomy, history *must* be learned. They cannot be replaced by magic, astrology, or
by a study of legends.[1]

49

R. W. Home (ed.), Science Under Scrutiny, 49–79.
© *1983 by D. Reidel Publishing Company.*

Feyerabend is not the only one asking, 'What's so great about science?'. Listen, for example, to Theodore Roszak:

... there is something radically and systematically wrong with our culture ... which frustrates our best efforts to achieve wholeness. I am convinced that it is our ingrained commitment to the scientific picture of nature that hangs us up.

What *is* to blame is the root assumption that culture – if it is to be cleansed of superstition and reclaimed for humanitarian values – must be wholly entrusted to the mindscape of scientific rationality.[2]

We are told that the 'scientific picture of nature' is only one picture, certainly no better and probably a lot worse than other pictures such as the 'magical picture'. Sometimes the point is put in terms of 'separate realities', 'scientific reality' being the most superficial. As for 'scientific rationality', it is only one form of rationality and a narrow and constraining one at that. Views like this are popular, and help to explain the 'flight from science' which gives some people sleepless nights. Feyerabend's question 'What's so great about science?' deserves an answer from those who study science.

The question is, first and foremost, an evaluative question.[3] The positive answer to it that I favour is an old one, easy to state but unexpectedly difficult to defend. It says that what is great about science is its ability to make intellectual progress; and what is great about 'scientific rationality' or 'scientific method' is that it fosters such progress. This is really only the beginnings of an answer. We can speak of progress only relative to some aim or aims. So to answer the question more fully, we would have to state and argue for some view about the aim or aims of science. For progress to occur, what comes later must be better than what precedes it. So we would need to state and argue for the criteria or values by which we think developments in science should be appraised, relating these to the aims of science. And this would lead us to most of the special problems of the philosophy of science, concerning explanatory power, evidential support, simplicity, fertility, and the like.

Having answered these questions, we could ask whether science has in fact made any progress. Obviously, the fact that science has evolved in a certain way does not entail that it has progressed, that later science is better than earlier science. Evolutionary ethics is as fallacious here as anywhere else. Obviously, scientific change can be described as progress only if we employ, implicitly or explicitly, some normative or evaluative view. Equally obviously, we can only ask whether scientific change has been progressive *after* we have found out how science has changed. So the problem of scientific progress is a two-fold one: What is scientific progress? and, Has scientific progress occurred?

Now as I have said already, trying to give a positive answer to these questions would be one way of answering the critics of science. It would not be an easy way, for they are not easy questions. Nor would any positive solution to the problem convince Paul Feyerabend. On being told 'What's great about science is that it has made intellectual progress', Feyerabend will ask immediately, 'What's so great about intellectual progress?'. And on being told, say, 'But that means that scientists have discovered more and more truth about the world', Feyerabend will ask immediately, 'What's so great about truth?'.[4] But Feyerabend apart, a positive solution to the problem of scientific progress would be of some value.

I am not going to discuss any such solution in this paper. Instead, I want to try to remove an obstacle that lies in the way of *any* adequate solution. The obstacle is an unholy alliance between facts and values which infects much modern thinking about science. Notoriously, recent philosophy of science has become 'historical' and recent history of science (or at least, some of it) has become philosophical. The watchword is, 'Philosophy of science without history of science is empty; history of science without philosophy of science is blind', a Kantian epigram which, since Imre Lakatos first coined it twenty years ago, has become almost as famous as its original. I will be asking to what extent the epigram is *true*. I will argue that the mixing of historical facts and philosophical values which it often engenders prevents us giving an adequate answer to the problem of scientific progress; and that this has, in turn, lent support to the view that there is nothing very special about science.

Philosophy has very few solid discoveries to its credit, but it does have one or two. The dichotomy between facts and values is one. It says that a logical gulf separates factual judgments and value-judgments: from purely factual premises no value judgment can be validly derived, and from value judgments no factual statement can be derived. I will be taking the fact/value dichotomy for granted.[5]

Now if we approach some contemporary studies of science armed with this dichotomy, we are bound to be perplexed. There is talk of a 'quantitative valuation and intercomparison of scientific activity, productivity *and progress*' (to quote the editorial statement in the first issue of a recently founded journal, *Scientometrics*[6]). Let me pause to clear this particular intellectual aberration out of the way.

Sociologists are keen to be scientific, which means, among other things, being 'objective' and 'precise'. Some sociologists of science think that 'objectivity' demands that they take no account of the scientific content of the activities they study. And they think that 'precision' demands that they count things (after all, if you don't count, you won't count). So they count lots of things scientific, and all the counts reveal a remarkable increase in scientific activity. More scientists are now alive than dead, they are fond of telling us. Scientific 'productivity', measured by the number of scientific papers published (or some subtle variant thereof), increases exponentially. So does the number of scientific journals in which these publications appear. On all these counts, 'little science' has grown into 'BIG SCIENCE'.

None of this shows that science has made intellectual progress. Nobody in their right mind would want to identify scientific progress with a growth in the number of scientists, or their publications, or their journals. (In so far as sociologists are tempted into such blatant violations of the fact/value dichotomy, they are out of their minds.) It is not the aim of science to provide a good living for more and more scientists, or ever more outlets for their ever increasing outputs. The growth of scientific knowledge cannot be identified with the growth in the number of scientific publications (not even if we weight each publication by the number of times it is referred to in other publications, which is a favourite 'objective' way of judging 'scientific quality'[7]). The question remains whether 'BIG SCIENCE' is good science, and 'BIGGER SCIENCE' better science. We need a philosophical theory of good science, however 'difficult to operationalise' it may be[8], before we can evaluate the growth of scientific activity at all.

No mere description of the evolution of 'little science' into 'BIG SCIENCE' can satisfy the critics of science. For them, this evolution is part of the problem. They see 'BIG SCIENCE' as an expensive game whose players have a vested interest in its continuing. They ask why tax-payers' money should be poured into science, and not into astrology, Velikovskian astronomy, or Black Magic. (Actually, what David Stove calls 'modern nervousness' is so widespread that quite a bit of tax-payers' money is poured into some of these things.) They ask why the wishes of the scientific community should be taken so seriously, to the detriment of the wishes of the community of astrologers or practitioners of the Black Arts. These are all serious questions. 'Quantitative Studies of Science', whatever their value, can do nothing to answer them.

It is less easy to deal with some other apparent violations of the fact/value dichotomy. There is talk of an irreducibly historical and/or sociological account of scientific progress. But how can this be, if history and sociology issue in factual propositions while talk of scientific progress is evaluative? Again, there is talk of confirming or refuting normative views about science by appeal to historical facts; and there is talk of using normative views to explain historical developments in science. But how can normative views be confirmed or refuted by historical facts, or explain them? Viewed from the perspective of the fact/value dichotomy, all this is a logical mess. I will try to clean up the mess, and to clarify the relationships between factual questions about science (which I take to be the province of history, psychology, and sociology of science), and normative or evaluative questions about science (which I take to be the province of philosophy of science).

Here a disclaimer is called for. I am not an academic trade-unionist seeking to lay down lines of demarcation. In saying that evaluative questions about science are the province of philosophy of science, I am not saying that only professional philosophers can ask and answer them. That would be a ridiculous thing to say. After all, it is scientists, and not philosophers, who evaluate their own and others' work — and it is their decisions, and not the decisions of philosophers, which counted in the past and will presumably continue to count. Most if not all of the interesting evaluative questions about science were first propounded by scientists. Notoriously, however, scientists have often disagreed with one another, both over particular evaluations and over general evaluative principles. Normative philosophy of science is simply a critical investigation of these often-conflicting principles, and of the arguments which can be produced for and against them.[9] And to repeat, I do not think that only professional philosophers can engage in the enterprise.

Nor must history and sociology of science be reserved for historians or sociologists. Scientists or philosophers might well ask and answer historical or sociological questions about science. Whether their answers are any good is to be determined by inspecting the answers, not by inspecting the 'professional affiliations' of those who produced them. When Pearce Williams asks 'Should Philosophers of Science Be Allowed to Write History?' and answers with 'a resounding "NO!"', he professes exactly the sort of view I am here disclaiming.[10] Such academic trade-unionism already involves a violation of the fact/value dichotomy, in so far as sociological facts about the producer of a view are taken to settle evaluative questions about the merits of the view produced.

2. GETTING VALUES FROM FACTS

(a) *Historical methodologies*

With this proviso in mind, let me turn to the question of how a normative philosophy of science is to be obtained and evaluated. The most obvious feature of recent philosophy of science is the attempt to base philosophy of science upon a close study of historical and/or sociological features of science.

There are several reasons for the popularity of this view. The dream, shared by Bacon and Descartes, of producing by *a priori* speculation a fool-proof 'logic of scientific discovery' proved to be an illusion. Logical positivists switched attention from the 'context of discovery' to the 'context of justi-fication', and identified philosophy of science with the logical analysis of science. But this came to seem increasingly remote from the facts of scientific life, 'dried-up petty-foggery', as Einstein once called it. Philosophy of science had to get closer to real, living, growing science — but how?

One way was suggested by Wittgenstein's *Philosophical Investigations.* Just as the positivist had drawn inspiration from the *Tractatus,* so also philos-ophers of science like Watson, Toulmin and Hanson drew inspiration from the *Investigations.* According to the *Investigations,* philosophy is not the investi-gation of *the* logical structure of language, for there is no single 'language' with a fixed 'logical structure'. Instead, there are many ways of using language, or 'language games', which characterise different 'forms of life'. The task of the philosopher is to depict these 'forms of life', to chart the 'logical geography' of the 'concepts' peculiar to each of them, the sorts of reasoning appropriate in each, the questions that can be asked and answered and the questions that cannot (the so-called 'limiting questions').

Thus Stephen Toulmin castigates the 'formal logician' for seeking a 'topic neutral' logic which will pronounce on the validity or invalidity of arguments from any walk of life:

What has to be recognised first is that validity is an intra-field, not an inter-field notion. Arguments within any field can be judged by standards appropriate within that field, and some will fall short; but it must be expected that the standards will be field-dependent, and that the merits to be demanded of an argument in one field will be found to be absent (in the nature of things) from entirely meritorious arguments in another.[11]

Hence for Toulmin there is no single logic and no single notion of logical validity. There is mathematical logic and mathematical validity, legal logic and legal validity, religious logic and religious validity, Azande magical logic and magical validity — and also scientific logic and scientific validity. And how is the philosopher (of mathematics, or law, or religion, or magic — or science) to *discover* the logical standards specific to his field? He will do it by an empirical and historical investigation which will

. . . study the ways of arguing which have established themselves in any sphere, accepting them as historical facts; knowing that they may be superseded, but only as the result of a revolutionary advance in our methods of thought. In some cases our methods will not be further justifiable — at any rate by argument: the fact that they have established themselves in practice may have to be enough for us. (In these cases the propriety of our intellectual methods will be what the late R. G. Collingwood called an 'absolute presupposition'.) [12]

By now the nature of Wittgensteinian philosophy of science should be clear. It will be a fundamentally *descriptive* enterprise. It will depict the 'ways of arguing' which have 'established themselves' in that particular 'form of life' and the standards which are there deemed appropriate. It must not be assumed, of course, that these scientific standards and ways of arguing are always and everywhere the same: the Aristotelian 'form of life' will be different from the Newtonian, the physicist's 'form of life' different from the biologist's. The Wittgensteinian philosopher of science must chart all these changes and differences. It will be a big task: hence Toulmin's *Human Understanding* is a big book. As Imre Lakatos once joked, for Toulmin a full-stop is a 'systematically misleading expression'.[13]

I have not time to argue against all this, not even against the fundamental misconception of the nature of logic on which it rests. Psychologism was the thesis, long ago refuted, that logic describes the way in which we think. Locke's 'new way of ideas' has given way to a new way of words, and we now have the thesis that logic describes the multifarious ways in which we talk. Pouring old psychologistic wine into new linguistic bottles has not improved its flavour: both views overlook the logician's prime concern with validity and invalidity. (I have logicians about me to *evaluate* my arguments — not to watch how I argue and then to devise a 'deviant logic' to accommodate me.) But as I say, this is not yet an argument, and I will not pause to develop it into one.[14]

Instead, I will merely ask whether the Wittgensteinian philosopher of science can say anything to the radical critics of science. It is clear that he cannot. For him, questions like 'What's so great about science?' or 'Is the scientific "picture of nature" better than the magical picture?' are silly questions, 'limiting questions' which transcend all 'forms of life' and so do not really make sense. Science exists, and so (still) does magic. They are very different 'forms of life', with different 'ways of arguing' which lead to different 'pictures of nature'. The philosopher can depict the differences, but he cannot presume to comment upon the superiority of one 'form of life' over another. Philosophy, as the master put it, leaves everything as it is. In dismissing questions like 'Is the scientific "form of life" any better than the magical?' as pseudo-questions, Wittgenstein's philosophy is an important source of contemporary philosophical relativism.

No mere description of what goes in science can say anything about scientific progress or about the value of science. With that in mind, let us now turn to the work which really put the descriptivist view on the map, Thomas Kuhn's *Structure of Scientific Revolutions*. (People often find it odd that Kuhn's book, apparently so destructive of logical positivist orthodoxy, appeared in the positivist *Encyclopaedia of Unified Science*. But this is not really odd. According to positivist dogma, a meaningful philosophy of science must be either logical analysis of science or descriptive socio-psychology of scientific communities. Orthodox positivists explored the first possibility, Kuhn the second.)

Kuhn's discussion of scientific progress begins, as had Popper's, with the declaration that progress is an 'obvious attribute' of the sciences, indeed, that it is 'reserved almost exclusively for the activities we call science'.[15] We might expect Kuhn to tell us next what scientific progress is, and then to show that it has occurred. But instead he warns us that his account will involve 'an inversion of our normal view of the relation between scientific activity and the community that practises it', and that 'the phrases "scientific progress" and even "scientific objectivity" may come to seem in part redundant'.[16]

Kuhn then gives us a socio-political account of scientific revolutions. The victorious party will always *say* that they have made progress, and will indoctrinate new practitioners accordingly. Hence it is 'not altogether inappropriate' to describe these new practitioners as victims of a 1984-style rewriting of history: it is 'not entirely wrong' to say that in the sciences, 'Might makes Right'.[17] Yet it is not entirely correct either. A revolution will not be a proper scientific revolution unless it is effected by a professional community of scientists, and not by any non-professional external authority. Professional communities are 'the sole possessors of the rules of the [scientific] game or of some equivalent basis for unequivocal judgements'.[18] They are the ones who should make the decisions, for 'What better criterion than the decision of the scientific group could there be?'.[19] Might does not necessarily make right; only professional might does.

All this aroused much controversy. Kuhn was accused of defending irrationalism, relativism, and 'mob rule'.[20] He professed to be surprised and puzzled by these accusations.[21] And his subsequent accounts of scientific change seem to bring him closer to his critics. I will argue, however, that a fundamental gulf still separates him from them.

Kuhn now emphasises that scientists choose between theories in the light of scientific values, conceding that he had not given enough attention to this in his original account.[22] He also concedes that his list of scientific values is a pretty familiar one: accuracy, consistency, scope, simplicity, and fruitfulness.[23] Finally, notwithstanding his earlier statements to the contrary, he concedes that values like these are not 'paradigm-dependent' but instead are relatively permanent features of science:

If the list of relevant values is kept short (I have mentioned five, not all independent) and if their specification is left vague, then such values as accuracy, scope, and fruitfulness are permanent attributes of science.[24]

Kuhn does insist on two points which may not be so familiar. First, that each of these values is difficult to apply, so that there can be genuine room for disagreement about their verdict in concrete cases. Second, that different values may give different verdicts (one theory may be simpler than another, but less accurate), so that the combined verdict of all of them may be equivocal. This means, Kuhn says, that in matters of theory-choice there is nothing like a logico-mathematical proof to compel the assent of all reasonable men. This does not make theory-choice an irrational matter of taste or a matter for mob psychology: it merely makes it judgemental, so that reasonable men can reasonably disagree in particular cases.[25]

Despite all this, a worry remains. It is located in passages such as these:

. . . the explanation [of scientific progress] must, in the final analysis, be psychological or sociological. It must, that is, be a description of a value system, an ideology, together with an analysis of the institutions through which that system is transmitted and enforced. Knowing what scientists value, we may hope to understand what problems they will undertake and what choices they will make in particular circumstances of conflict. I doubt there is another sort of answer to be found.

Some of the principles deployed in my explanation of science are irreducibly sociological, at least at this time. . . . Whatever scientific progress may be, we must account for it by examining the nature of the scientific group, discovering what it values, what it tolerates, what it disdains.

If I sometimes say that any choice made by scientists on the basis of their past experience and in conformity with their traditional values is *ipso facto* valid science for its time, I am only underscoring a tautology. Decisions made in other ways or decisions that could not be made in this way provide no basis for science and would not be scientific.[26]

Now if we describe an ideology or value-system and the way it is transmitted and enforced, we do not leave the realm of social psychology. (It is perhaps worth stressing that 'Scientists value precision' is a factual statement, not an evaluative one.)[27] And if we explain historical developments with reference to this ideology, we do not leave the realm of history of science. All this leaves untouched the question of whether the ideology is a defensible one and whether the changes it helps to explain represent progress. Kuhn's verbal legislation, that decisions made in other ways would not be scientific decisions, does not help either. Radical critics of science will not be upset by being told that they want something other than science: that, they might say, is precisely their point.

This residual worry can easily be resolved in the following way. Kuhn's socio-psychological descriptions can be transformed into normative views if we suppose that he endorses the norms and values he finds in science, saying, for example, not merely that scientists value precision but that precision is valuable. And in defence of this approach, we could argue that a philosophy which pays no heed to the things scientists value would hardly be a philosophy of *science* at all.

Such an approach has two limitations, however. First, and most important, by itself it yields no good philosophical argument for the values it contains. Of course, the *non sequitur* 'Scientists value precision: therefore, precision is valuable' can easily be repaired by adding the extra premise, 'Scientific might makes right: whatever scientists value is *ipso facto* valuable'. But this is to dodge the problem, not to answer it: doubts about the original *non sequitur* are transformed into precisely similar doubts about the extra premise which repairs it. We might discover that scientists value simplicity and tentatively endorse this value; but the philosophical question remains of *why* a simpler theory is better than a less simple one. We might discover that scientists value theories which make successful new predictions, and tentatively endorse that value: but the philosophical question remains of *why* such theories are preferable to those which merely explain known facts. We might discover that scientists value precision and tentatively endorse that value; but the philosophical question remains of *why* precision should be valued. Socio-psychological investigation of scientific communities, followed by a mere endorsement of the values we find there, yields no philosophical defence of those values and so cannot answer the question, 'What's so great about science?'.[28]

The second limitation of this approach lies in its tacit assumption that socio-psychological investigation will reveal that scientists have a consistent set of values. What if scientists are found to disagree about the criteria by which contributions to science are to be assessed? The philosopher could not then endorse *any* value to which he finds scientists appealing, on pain of contradiction. The philosopher would have to examine these competing values before he could even endorse a consistent set of them. Alternatively, of course, he could resign himself to a relativistic position, and say that different 'communities' have different values and that there is nothing to choose between them. But if Newtonian physics is only better than Aristotelian physics by Newtonian standards, while the reverse is the case by Aristotelian standards, we hardly have anything like scientific progress. And we hardly have any way of answering those who say that Azande magic is better than either Aristotelian or Newtonian physics, by Azande standards.

Now as we have seen, Kuhn avoids this second limitation by claiming that certain scientific values are 'permanent attributes of science'. But Laudan (and Toulmin) disagree. According to Laudan,

... such components of rational appraisal as criteria of explanation, views about scientific testing, beliefs about the methods of inductive inference and the like have undergone enormous transformations.[29]

Laudan does not now examine these changing evalutive views and tell us which he prefers and why. Instead, he opts for historical relativism.

Laudan sees science as a problem-solving enterprise, as many have done before him. The only absolute evaluative principle he defends says that if one theory is a better problem-solver than another, then it is preferable. But, he insists, scientists' views about what counts as a genuine problem and what counts as an adequate solution have changed drastically over time. Laudan incorporates all of these changing views into his own philosophy. Hence when he appraises past theory-choices to determine whether they were progressive, he does so using the norms or values of the scientists who made the choices, while conceding that these may be very different from current norms or values.[30] This is an almost total endorsement of the view that Scientific Might makes Right. (The trivial exception is that Laudan will condemn a scientist who prefers a theory which is not the best problem-solver by that scientist's own lights.)

Thus Laudan arrives at a view almost indistinguishable from the Wittgensteinian one. Both maintain that philosophy of science is a quasi-empirical investigation of the 'form of life' we call 'science'. The Wittgensteinian merely records shifting 'patterns of argument' and 'standards'; Laudan makes each of them his own. (The difference is a subtle one: it is analogous to the difference between the factual statement, 'Nineteenth-century Englishmen believed that slavery is wrong', and the relativistic value-judgement, 'Slavery was wrong for nineteenth-century Englishmen'.) In neither case is there a philosophical defence of any of the principles recorded.[31] How *could* Laudan defend any of them, if they are such a mixed bunch?[32] Indeed, he does not even state any of them very precisely, presumably because they are yet to be thrown up by historico-sociological research.

Like all relativistic positions, Laudan's is powerless to prevent the relativism spreading. Laudan thinks Greek scientists had standards very different from those of seventeenth-century scientists. But did not competing schools of seventeenth-century scientists also have different standards? Did not Isaac Newton have standards different from those of Robert Hooke? If historical research answers these questions in the affirmative, then Laudan's historical relativism will degenerate into sociological relativism and even personal relativism.

Furthermore, once Laudan extends his enquiry beyond science, he is likely to find that astrologers and magicians have different standards from scientists. Consistency will compel him to endorse *those* standards in determining whether astrology and magic are 'rational and progressive', and no doubt the answer will be 'yes'. If we are not to compare competing scientific standards, then we cannot compare scientific standards with non-scientific ones. Once again, the question 'What's so great about science?' cannot be answered.

What leads Laudan to this position? What leads him there is a laudable desire for historical veracity combined with a mix-up of facts and values, or rather of factual problems (understanding or explaining a past scientific decision) and evaluative problems (appraising the reasonableness of that decision). If standards or values have changed, then obviously past scientific decisions will have been based on standards different from those of contemporary scientists. Historians will rightly be impatient with those who seek to *understand* past decisions in terms of current standards. For as Laudan says:

Unencumbered by modern notions of rationality, scientists of the past had to make decisions about the acceptability of contemporary theories by their criteria rather than by ours.[33]

Quite so. But understanding a past decision is one problem (a factual problem), evaluating it is another (a normative problem). Obviously, if we wish to understand or explain a past scientific decision, then we must invoke the standards of the time since only these can have influenced it. But if we wish to *evaluate* the decision, why should we not employ standards of our own, standards which we may think (with good reason) better than the past ones?

There is a well-known distinction between two senses in which an action or decision can be deemed rational. An action or decision is *rational in the weak sense* if it conforms to the beliefs and values of the agent. Thus, to take a trivial example, if a person believes that stepping on cracks in the pavement brings bad luck, then it is reasonable (in this weak sense) for him to avoid stepping on cracks. An action or decision is *rational in the strong sense* if it conforms to beliefs and values which are themselves reasonable ones. Thus, to pursue the trivial example, avoiding cracks in the pavement may be deemed irrational (in this strong sense) because the belief that stepping on cracks brings bad luck is itself unreasonable.[34] That a past scientific decision conforms to beliefs and values operative at the time shows only that it was rational in the weak sense. Someone who thinks past values unsatisfactory may still wish to deem the decision an unreasonable one. But Laudan, by weakening his methodology so that it incorporates whatever standards happen to have been operative in the past, is unable to do this.

Suppose, to vary Laudan's own example, that once upon a time an accepted criterion for evaluating any theory was to ask whether it was compatible with the Scriptures, literally interpreted. Suppose some geological hypothesis was correctly deemed incompatible with the Scriptures, and rejected on this ground. This decision could be explained, and shown to be rational in the weak sense, by reference to the criterion. But it remains an open question whether it was a rational decision in the strong sense, that is, whether compatibility with the Scriptures is an acceptable criterion of scientific merit. On Laudan's relativistic view, this is not an open question at all.

So far I have been granting Laudan's claim that there have been enormous changes in scientific methods and values. Is this really so? Some of Laudan's own examples of changes in standards are problematic.[35] And we must distinguish standards appealed to by scientists in their philosophical moments from standards applied by them in practice; I suspect that there has been more variation in the former than in the latter. But even if standards have changed, and may change again, it still makes perfectly good sense to ask whether the changes were, and are, for the better.

(b) *Testing methodologies*

So far I have resisted the idea that a normative philosophy of science can be *obtained* simply from historical and/or sociological investigations into science. Now I turn to the question of whether a normative philosophy of science can itself be *evaluated* by appeal to historical and/or sociological facts. That is, can the history of science adjudicate between rival philosophies of science?

Can one criticise a normative methodology by showing that scientists have not in fact conformed to it?

Viewed from the perspective of the fact/value dichotomy, such criticism is misconceived. It proceeds from a factual statement to the negation of an evaluative statement. But since the negation of an evaluative statement is itself evaluative, this is to proceed from facts to values. It is no criticism of 'Thou shalt not lie' to point out that people sometimes tell lies. And it is no criticism of 'Thou shalt not propose *ad hoc* hypotheses' to point out that scientists do sometimes propose *ad hoc* hypotheses. The logical situation is not improved by the additional fact that the liar (or the scientist who indulged in *ad hoc*ery) was a great man or an important historical figure. 'George Washington told a lie' fails to contradict 'Thou shalt not lie'. And 'Galileo proposed an *ad hoc* hypothesis' fails to contradict 'Thou shalt not propose *ad hoc* hypotheses'.

Such *non sequiturs* can easily be repaired. I can criticise 'Thou shalt not lie' by pointing to a particular case where a person rightly told a lie. Here I use a particular normative judgement to refute a general normative rule, and no violation of the fact/value dichotomy is involved. The procedure has an obvious rationale. We may have competing systems of moral generalisations (not to mention competing systems of meta-ethics), while agreeing on the rights and wrongs of particular cases. When this is so, moral judgements about particular cases can be used as a testing-ground for competing moral systems. This is precisely the way in which we do criticize low-level moral theories like 'Thou shalt not lie', as well as high-level theories like utilitarianism.

Imre Lakatos and Larry Laudan propose that a normative philosophy of science can be criticised in a similar way. The rationale of the proposal is again obvious. Francis Bacon could condemn the science of his day, because there was not much of it. But nowadays there are one or two universally acknowledged scientific achievements. Scientists are agreed that Galileo's theory of terrestrial motion was better than Aristotle's, Newton's astronomy better than Descartes', Lavoisier's theory of combustion better than the phlogiston theory. Agreement on these statements does not, of course, entail agreement on the reasons why they are true. Lakatos and Laudan propose that any philosophy of science should try to explain such fundamental value judgements, to say why they hold. And it will be a *prima facie* objection to a philosophy of science if it fails to explain one of them. (But only a *prima facie* objection: as with experimental reports in science and particular moral judgements in ethics, these particular value judgements are not sacrosanct, and might upon reflection be rejected.)

This, then, is the Lakatos-Laudan proposal — or as much of it as I am prepared to accept. A philosopher who did not at least try to explain the fundamental judgements I just listed, and a few others like them, would not be doing philosophy of *science* at all. The proposal is not that we can refute methodological values by historical facts, but rather that we can refute a general system of values by particular value judgements. It is not that 'philosophy of science without history of science is empty', as the epigram has it. It is rather that a philosophy of science can be criticised by appealing to certain fundamental value judgements about the historical development of science.

It might be objected that this proposal begs the question of scientific progress. For in demanding that a philosophy of science explain why Galileo's theory is better than Aristotle's, do we not ensure that science will have made progress in this case (and a few others)? I do not think so. For the proposal does not guarantee that any adequate philosophy of science can be produced. The proposal says merely that if science has progressed at all, then it has done so in the Aristotle-Galileo case (and a few others). To propose an adequacy-condition for philosophies of science is not to guarantee that any adequate philosophy can be produced.

However, Lakatos seems to have extended his original proposal into one which I find unacceptable. The extended proposal is that a philosophy of science should seek to explain not just a few particular value judgements, but any value judgement agreed upon by scientists of the basic form, 'At time *t* theory *A* was better than theory *B*'. Lakatos says that a methodology should try 'to interpret more of the *actual* basic value judgements in the history of science as rational', so that

> ... *progress in the theory of scientific rationality is marked* ... *by the reconstruction of a growing bulk of value-impregnated history as rational.*[36]

Now as these remarks suggest, Lakatos does not think that any methodology will completely succeed: there will always be anomalies, basic value judgements made by scientists which contradict the dictates of the methodology. Here Lakatos introduces a distinction between 'internal' and 'external' history of science. Internal history of science explains past scientific decisions which conform to the methodology (I will be discussing this in a moment). External history of science explains the anomalies, by tracing them to 'external factors'. The better the methodology, the more past decisions are interpreted by it as rational ones.[37] (Lakatos himself stresses that his internal/external distinction is methodology-dependent: an 'external factor' for a methodology is anything not mentioned in that methodology, and what is 'external' for one methodology may be 'internal' for a different one.)

This extended proposal evoked much criticism. It was objected that the best methodology of all, on this extended adequacy condition, would be the single principle, 'Whatever scientists agree upon is right'. It was objected that Lakatos too was committed to the idea that Scientific Might makes Right, or to the Hegelian idea that Whatever is Real is Rational. And it was objected that this extended adequacy-condition for methodologies deprives the philosopher of science of any critical function (and, we might add, of any means of answering the radical critics of science).[38]

John Worrall tries to rebut such criticisms. He begins by pointing out, rightly, that:

... the naturalistic fallacy being genuine, [a] methodology that tells us what scientists *ought* to prefer will not be directly refuted if scientists' actual preferences are different.[39]

Worrall leaps the is-ought gap by suggesting that any methodology should be taken to make the historical claim that the preferences of scientists conform to the methodology *unless* they are influenced by 'external factors'. A methodology (I would say, this historical claim) will be historically confirmed by cases where the preferences of scientists conformed to the methodology and by cases where they did not but where there is historical evidence of the operation of an 'external factor'. A methodology (I would say, its associated historical claim) will be historically refuted by cases where the preferences of scientists violated the methodology, yet there is no evidence of the operation of an 'external factor'. Worrall emphasises that this is *not* to say that a methodology is the better, the more past decisions it counts as rational:

On my account, on the contrary, one methodology may be better confirmed than another even if it explains fewer historical developments 'internally'. For a methodology may be confirmed by historical cases of which it gives an *external* explanation, *provided* it can give independent historical evidence for the existence of the external factors it invokes.[40]

But this extended proposal for testing methodologies against the history of science is either too weak or too strong. Let us suppose that the actual preferences of scientists are brought about somehow, that is, are historically explicable (a not unreasonable supposition). And let us recall that for Worrall, as for Lakatos, an 'external factor' for a methodology is anything not mentioned in that methodology. It follows that no matter what our methodology, any actual decision will be brought about either by internal factors, thus confirming the methodology, or by external factors for which we could presumably find evidence, thus confirming the methodology again.[41] The only past decisions which could refute a methodology would be decisions brought about by nothing at all – not even by tossing a coin, for that would presumably be an 'external factor' too. But decisions of that kind are logically impossible. And hence the proposal is too weak; in fact, it is empty.

We can give the proposal some teeth – but if we do, they will be too sharp. We can drop the methodology-dependent definition of 'external factors' and revert to the traditional, if vague, definition of 'external factor' as something which has nothing to do with reasons of any kind.[42] But now suppose an adherent of methodology M finds that a past decision was based, not on external factors in this traditional sense, but on a rival methodology $M*$. Why should it be an objection to methodology M that it deems that decision unreasonable, because it was based on the unreasonable methodology $M*$? Why should a philosopher of science deem all the reasoned decisions of scientists reasonable ones? Thus construed, Worrall's proposal is too strong. It marks, once again, a return to the view that scientific might makes right, that whatever scientists do for a reason is *ipso facto* reasonable.

And having returned to that view, *via* an unacceptably strong adequacy condition on any philosophy of science, we once again prevent philosophy giving any convincing reply to the critics of science. These critics will not be satisfied by mere endorsements, either of some of the values of scientists, or of some of the decisions they have reached by applying those values. They want philosophical arguments for the values and the decisions to which they lead. But 'historical methodology' by itself cannot give such arguments.

So far I have been defending the integrity of philosophy of science against the encroachments of historicism and sociologism. Now I want to defend the intetegrity of history of science, and even of sociology of science, against the encroachments of philosophy. Specifically, I want to turn to the idea of 'rationally reconstructed' history of science – 'internal history', as Lakatos calls it – and the subtle derivation of facts from values which it contains.

3. OBTAINING FACTS FROM VALUES

Once we have normative philosophy of science, the interesting question arises of the extent to which actual science has been as the philosophy says it ought to have been. In particular, once we have an account of when one scientific theory is preferable to another, we can ask whether the actual preferences of scientists have followed the dictates of the methodology. Lakatos proposes that the history of science be written with such questions in mind, and in the following way. Confronted with some historical episode, you first say what ought to have happened according to your philosophy: this is the 'internal history' or the 'rationally reconstructed history'. Then you inspect the historical record to see how well actual history conforms to the rational reconstruction. It will never conform perfectly, and so 'external history' will always be needed to explain the discrepancies. Lakatos writes:

The history of science is always richer than its rational reconstruction. *But rational reconstruction or internal history is primary, external history only secondary, since the most important problems of external history are defined by internal history.* External history either provides non-rational explanation of the speed, locality, selectiveness, etc. of historic events as *interpreted* in terms of internal history; or, when history differs from its rational reconstruction, it provides an empirical explanation of why it differs. But the *rational* aspect of scientific growth is fully accounted for by one's logic of scientific discovery.[43]

Lakatos goes on to suggest an interesting way of presenting a historical study. You relate the 'rationally reconstructed' history in the text, and indicate in the footnotes how actual history 'misbehaved'. Thus in Lakatos's own text we find statements about what Prout 'knew very well' and 'said' — and the footnote tells us that actually Prout thought and said the opposite! Or we find a statement about what Bohr 'thought' in 1913 — and a footnote telling us that he only thought this in 1926![44]

Everybody is horrified. Kuhn says that this 'is not history at all but philosophy fabricating examples'. Holton calls it 'an ahistorical parody that makes one's hair stand on end'. McMullin says that what results from this 'extraordinary procedure' is 'not history'. Laudan says that it is 'consciously and deliberately falsifying the historical record'.[45]

Some of the hysteria is, I think, misplaced. It is hardly a falsification of the historical record to set that record straight, albeit in a footnote. Lakatos does not merely tell us what ought to have happened in these cases, as McMullin and Laudan claim; he also tells us what did happen. The two are not always the same, and orthodox historians quite often draw attention to the fact. In these cases, Lakatos merely suggests a colourful way of doing something quite orthodox.[46]

There is a more fundamental objection. It concerns cases where history *fails* to misbehave, where 'internal history' does *not* have to be supplemented by 'external history'. Lakatos tells us that 'internal or rationally reconstructed' history

... takes place essentially in the world of ideas, in Plato's and Popper's 'third world', in the world of articulated knowledge which is independent of knowing subjects.

. . . is not dependent in the slightest on the scientists' beliefs, personalities or authority. These subjective factors are of no interest for any internal history.

... will not need to take any interest whatsoever in the *persons* involved, or in their beliefs about their own activities.[47]

So when history fails to misbehave, we need make no mention of the beliefs or values of the historical agents involved. When scientists do what the philosopher thinks they ought to have done, then the philosopher-historian can explain what happened simply by citing his view about what ought to have happened. In these cases, as Lakatos puts it, 'scientific growth is *fully accounted for* by one's logic of scientific discovery'. But this is obviously to derive facts from values.[48]

We cannot explain a past decision merely by pointing out that it was the right decision according to some methodology, if only because people can do the right thing for the wrong reasons. We must add the historical claim that the past decision was actually based upon the methodology in question.[49] Notice that this will be a claim about the beliefs and values of historical agents, things which were supposed not to figure in 'internal history'. Notice too that once we do add this historical claim, then any normative philosophical thesis becomes a redundant premise in our historical explanation. We have explained the past decision entirely in terms of the beliefs and values of the people who made that decision (whether the explanation is true is another question, of course). And this is just as it should be.[50]

Lakatos (and Worrall) later modified their position — but not enough. Lakatos conceded that 'an appraisal criterion alone cannot possibly explain the actual history of science'. An appraisal criterion yields statements of the form 'At time t theory A was superior to theory B'. We must also, Lakatos said, add 'psychological premises' to the effect that 'All scientists will — *ceteris paribus* — accept A over B at time t if A is superior to B at time t' and 'in this case the *ceteris paribus* condition was satisfied'. Explanations of rational scientific change require both normative and factual ('psychological') assumptions. But the normative assumption is still necessary:

All historians of science who distinguish between progress and degeneration ... are bound to use a 'third world' premise of appraisal in explaining scientific change.[51]

This is still an unholy alliance of facts and values which needs to be severed. If we want to explain why scientists preferred T_1 to T_2, all we need to invoke are facts about those scientists themselves (including, perhaps, facts about which methodological values they applied). The statement that T_1 was *better* than T_2 (according to some normative philosophy of science) can play no role in any historical explanation.[52] Of course, having explained the past decision, we might also want to ask whether it was a reasonable one, that is, whether the change from T_1 to T_2 represented progress. And obviously, to answer this question we must invoke some theory of rationality, some normative theory about the circumstances in which one theory is better than another. Historians who distinguish between progress and degeneration must indeed make appraisals of what happened in history. But appraising what happened and explaining what happened are two different enterprises, which ought to be kept distinct. Lakatos's 'rational reconstructions' merge the two enterprises, and do a disservice to both.

All this stems, as Lakatos's references suggest, from Popper's views on the 'third world'. Popper distinguishes three 'worlds': the 'first world' of physical objects and processes; the 'second world' of mental states and processes; and the 'third world' of objective knowledge. And he claims that 'third-world' facts can causally influence the 'mental world' and, through that, the physical world, so that neither the physical world nor the mental world is 'causally-closed'. An example will make the idea clear. Suppose that some physical theory T logically implies C, where C is a recipe for making an atom bomb. This logical fact is a 'third-world' fact. Suppose further that scientists recognise this logical fact, construct the bomb, and then explode it, with obvious effects on the physical world. Popper claims that appeal to the 'third-world' fact that T logically implies C is essential if we are to explain the activities of scientists and the physical repercussions of their activities.

This, too, is wrong. What we need in order to explain the activities of scientists is not the statement 'T logically implies C', but rather the 'second-world' statement, 'Scientists believed that T logically implies C', or, better, 'Scientists derived C from T'. To explain what scientists think and do, we need refer only to the beliefs and values which inspire their activities. The 'third-world' logical fact is relevant only if we wish to *evaluate* the cogency of someone's logical beliefs or the validity of his arguments.[53]

So Kuhn was basically right: 'rationally reconstructed history' is not proper history at all. How could it be, when the beliefs and values of historical agents need not figure in it? It follows from the fact/value dichotomy that no factual explanation can contain non-redundant normative premises.

'Rationally reconstructed history' violates that dichotomy. It remains true that any normative methodology prompts the interesting question, 'To what extent has the history of science conformed to this methodology?' (Just as any morality prompts the interesting question, 'To what extent have the affairs of men conformed to this morality?') It also remains true that this is the crucial question for any philosopher who wishes to show that science has made intellectual progress (and thereby to answer the radical critics of science). But in answering the question, we need real history, not some 'rational reconstruction' of it.

But why ask the question at all? Why not simply describe and explain what has happened in the history of science, without also trying to evaluate it? This is certainly possible, notwithstanding some rather glib arguments to the contrary.[54] But such an enterprise will only depict scientific change; it cannot, without violating the fact/value dichotomy, say anything about scientific progress. History of science without philosophy of science is not blind, as the epigram has it — it is merely blind to scientific progress. And because the sciences seem to present the best example, perhaps even the only example, of sustained intellectual progress, historians have typically appraised what has happened as well as describing and explaining it. I am not opposed to 'norm-impregnated history' of this kind. Indeed, we need such historical investigations in order to answer the question, 'What's so great about science?' But if the answer is to be at all convincing, we must not conflate epistemological norms and historical facts, or forget that the norms must be articulated and defended by philosophical argument.

A last point. Our topic for this symposium was 'History and Philosophy of Science: Their Educational Role and Bearing on Contemporary Problems'. I began with what I regard as an important contemporary problem: the flowering of the 'counter-culture' and the flight from science, not to mention reason, which it inspires. So let me end with a word about science education. Suppose that what is great about science *is* its ability to make genuine intellectual progress. I have not shown this, of course; but suppose that it is true. Suppose, further, that Kuhn is right about orthodox science education: that it is based upon textbooks which are profoundly unhistorical, and profoundly uncritical to boot. It follows from these two suppositions that orthodox science education does not teach what is great about science. For obviously, if the progressive nature of science is to be taught, then past science must figure as well as current orthodoxy. Many problems stand in the way of a historical or even quasi-historical approach to science education, not least the exigencies of time. But the sort of textbook indoctrination described by Kuhn conveys neither a proper understanding of current science nor a proper appreciation of what is great about it.

University of Otago

NOTES

1 For Feyerabend's question, see his 'On the Critique of Scientific Reason', p. 310; for the complaint, see his *Against Method*, pp. 299 and 301.

2 'Some Thoughts on the Other Side of This Life', p. 45, and *Where the Wasteland Ends*, p. xxx; as quoted by Holton, *The Scientific Imagination*, p. 88. Holton answers the radical critics of science by showing, principally by quoting Albert Einstein (hardly a typical scientist!), that they are wrong about the way scientists proceed. He objects to attempts to construct a normative philosophy of science which might enable us to show that science has made intellectual progress. But the radical critics of science will not be pacified by accounts of the life and work of 'great scientists': they want to know what makes a scientist a great one, and why his work deserves more respect than that of a great astrologer or a great mystic.

3 Here, of course, I use 'evaluative' in an epistemological sense, not a moral one. Showing that science has made intellectual progress, that later science is epistemologically better than earlier science, leaves open the question of whether such progress is morally desirable.

4 John Watkins calls this 'effortless one-upmanship': see his 'Corroboration and the Problem of Content Comparison', p. 339 (and the references there given).

5 Philosophers are quarrelsome creatures, and some have quarrelled with the fact/value dichotomy. We can derive facts from values: 'John is a man' follows validly from 'John is a good man'. We can derive statements-containing the word 'ought' from factual statements: 'Snow is white' logically implies 'Either snow is white or one ought always to tell the truth'. Finally, since 'ought' implies 'can' (and 'not-can' implies 'not-ought'), any ought-statement implies a descriptive statement (and the negation of the descriptive statement implies the negation of the ought-statement).

We can see that the fact/value dichotomy must be carefully formulated if it is not to succumb to counter-examples like these. The following formulation will suffice for my purposes. Construe a value judgement as a statement of the form 'It ought to be the case that *p*' where *p* is some descriptive statement. (And, for example, construe value judgements like '*A* is preferable to *B*' as 'It ought to be the case that *A* is preferred to *B*'.) Then the fact/value dichotomy says that 'It ought to be the case that *p*' never follows from *p*, and that *p* never follows from 'It ought to be the case that *p*'. (Here I am indebted to R. G. Durrant.) As for the 'ought-can principle', I agree with James Brown ('Moral Theory and the Ought-Can Principle', p. 208) that this is itself a *moral* principle which figures as a suppressed premise in all alleged derivations of 'can-statements' from 'ought-statements' or of the negations of 'ought-statements' from the negations of 'can-statements'.

Finally, we must be careful not to confuse the fact/value dichotomy with the fact/theory dichotomy. Otherwise we will mistake, for example, arguments to the effect that factual propositions are often or always 'theory-laden' for arguments to the effect that they are often or always 'value-laden'. It is probably better to distinguish descriptive statements (including both factual and theoretical statements) from normative statements (including both epistemological and moral norms).

6 The editorial statement is by M. T. Beck, the italics are mine.

7 See, for example, G. Nigel Gilbert, 'Measuring the Growth of Science: A Review of Indicators of Scientific Growth', pp. 21–23.

8 Gilbert (*op. cit.*, p. 21) mentions the possibility of employing some philosophical criterion of scientific merit, but drops it because such criteria are 'difficult to operationalise'. Instead, we are to judge the quality of a scientific paper by seeing how many times it is cited in other papers, or by asking the compilers of bibliographies, or simply by canvassing scientific opinion. As Gilbert admits, such methods can at best only tell us what scientists think about scientific quality, not about scientific quality. (Sometimes they do not even do that: the most cited scientist for 1967 came out ahead of two future Nobel prize winners and even further ahead of a past Nobel prize winner (see Jevons, *Science Observed,* pp. 44–45); either the Nobel Prize Committee is failing to locate genuine scientific merit, or the Science Citation Index is!) Narin tells us that the 'most objective' method of measuring 'scientific *advance*' is simply to count up the number of publications, but that this is the 'least relevant' method. The 'most relevant' and 'most subjective' method is to ask eminent scientists whether science is advancing. Other methods fall upon a continuum between these two extremes. (See Narin, 'Objectivity versus Relevance in Studies of Scientific Advance'.) But the 'most relevant' method here can at best yield only 'Eminent scientists agree that science is advancing', a statement which is logically irrelevant to (does not imply) 'Science is advancing'.

9 Philosophy of science has also concerned itself traditionally with questions about the metaphysical presuppositions and/or implications of science (questions, for example, about space and time or causality and determinism). I will not be discussing such factual questions, though they are often connected with evaluative ones. (For a more detailed taxonomy of the various different concerns of philosophers of science, see Passmore's paper in the present volume.)

10 See Pearce Williams's article of this name; the 'resounding "No!"' occurs on p. 252. My objection is, of course, only to the academic trade-unionism implicit in Pearce Williams's *title*. In the course of his paper he gives substantive criticisms of two historical works which happen to have been written by philosophers, criticisms which must, of course, be judged on their merits.

11 Toulmin, *The Uses of Argument*, p. 255.

12 *Ibid.,* pp. 257–258.

13 Lakatos, 'Understanding Toulmin', p. 231.

4 A far from complete analysis of the demise of psychologism is given in my 'George Boole and Psychologism'. On Toulmin's views, see Jarvie, 'Toulmin and the Rationality of Science', and Briskman, 'Toulmin's Evolutionary Epistemology', as well as Lakatos, 'Understanding Toulmin'.

15 Kuhn, *Structure of Scientific Revolutions,* pp. 160–161 (see also the 'Postscript – 1969', p. 209). Popper remarks, in a similar context: '. . . in science . . . we can speak clearly and sensibly about making progress . . . . In most other fields of human endeavour there is change, but rarely progress (unless we adopt a very narrow view of our possible aims in life); for almost every gain is balanced, or more than balanced, by some

loss. And in most fields we do not even know how to evaluate change.' (*Conjectures and Refutations*, pp. 216–217)

[16] Kuhn, *Structure of Scientific Revolutions*, p. 162.

[17] *Ibid.*, p. 167.

[18] *Ibid.*, p. 168.

[19] *Ibid.*, p. 170. Compare Kuhn's later remark that '. . . trained scientists are, in such matters [of theory-choice], the highest court of appeal . . . ' (Reflections on My Critics', p. 234). Is this the uncontentious sociological thesis that (as things currently stand regarding scientific autonomy) trained scientists do in fact make the decisions? Or is it the highly contentious thesis that 'trained scientists' can never be wrong? Is it socio-logical description only, or philosophical prescription also? Kuhn tells us that it is both, because in his account 'the descriptive and the prescriptive are inextricably mixed' so that his 'remarks on scientific development' should be read both as descriptions and prescriptions ('Reflections on My Critics', pp. 233 and 237). So he does seem to be committed to the thesis that professionally organised might makes right.

[20] For example, by Shapere, 'Meaning and Scientific Change', p. 67, and by Lakatos, 'Falsification and the Methodology of Scientific Research Programmes', p. 178 (*Philosophical Papers*, I, 91).

[21] See, for example, his 'Reflections on My Critics', p. 260; *The Essential Tension*, p. 321.

[22] *Structure of Scientific Revolutions*, 'Postscript – 1969', p. 185.

[23] *The Essential Tension*, pp. 321–322.

[24] *Ibid.*, p. 335. Kuhn does caution us that it is only 'very roughly' the case that scien-tific values have not changed. There have been variations in the extent to which these permanent values have been applied and in the weight given to each. And 'practical utility' is a value in some sciences but not others. (One wonders how Kuhn would react to the claim that it ought to be a value in all sciences.) Kuhn also says that value changes are often consequent upon theory changes.

[25] The fullest discussion of these points is in Kuhn's *The Essential Tension*, Chapter 13. See also his *Structure of Scientific Revolutions*, 'Postscript – 1969', pp. 184–185 and 199–200; or 'Reflections on My Critics', p. 262.

[26] For these statements see 'Logic of Discovery or Psychology of Research', p. 21; 'Reflections on My Critics', pp. 237–238; *ibid.*, p. 263.

[27] This rather obvious point is overlooked by those who argue that the 'human sciences' cannot be value-free because their subject-matter, human beings, is 'value-laden'. But this is a *non sequitur* because descriptions of the values of human beings, and of the way these values influence their actions, are factual statements, not normative ones.

[28] Here I agree with Ernan McMullin and Ronald Giere. Giere writes: 'No one denies that scientists do in fact use various criteria to compare theories. The question is whether these are *rational* evaluations or merely means employed to persuade others to change their allegiances . . . . If one grants that epistemology is normative, it follows that one cannot get an epistemology out of the history of science – unless one provides a phil-osophical account which explains how norms are based on facts. This ought to be a central problem for historically oriented philosophers of science, but few seem willing even to acknowledge the question, let alone attack it head on.' ('History and Philosophy of Science: Intimate Relationship or Marriage of Convenience?', pp. 285 and 290.) McMullin calls 'methodology of science' a factual investigation into the criteria scientists

call upon to decide between theories. He then says: 'But suppose there is some uncertainty or even disagreement in their regard? Where does one turn then? The methodology of science is not self-certifying. If the question be carried to this second level, we can speak of a "metamethodology".' ('Philosophy of Science and Its Rational Reconstructions', pp. 223–224.) What McMullin calls 'methodology', I would call 'history cum sociology of scientists' values'; what he calls 'metamethodology', I would call 'normative philosophy of science'.

[29] *Progress and Its Problems*, p. 129. Lakatos agrees: 'Great works of art may change aesthetic standards – great scientific achievements may change scientific standards. The history of standards is the history of the critical – and not so critical – interaction between standards and achievements.' (*Philosophical Papers*, I, 201.) Despite this, Lakatos obviously thinks that we can critically examine changing standards and try to say which is better. Laudan, on the other hand, though he occasionally lets slip that he has views on this matter, incorporates all the changing standards into his own philosophy. (Lakatos's parallel between aesthetic standards and scientific ones may be misleading, for there is much to be said for the view that changes in aesthetic standards yields *incommensurable* aesthetic forms: we may well be able to determine whether one seventeenth-century landscape is better than another, but does it make much sense to say that an impressionistic landscape is better than both of them?)

[30] See *Progress and Its Problems*, esp. pp. 130–131.

[31] Laudan does occasionally remark that he thinks later scientific standards are *better* than earlier ones (see, for example, *Progress and Its Problems*, p. 129). But he gives no argument for these remarks. And it is always past standards which he uses to evaluate past scientific decisions, whether he knows of better ones or not.

[32] Laudan (like Kuhn before him) does close off one avenue along which one might seek to assess competing scientific norms: their efficacy in aiding the search for truth. Laudan says that science cannot be a search for truth because it can never achieve certainly. But this is a *non sequitur,* because truth and certainly are not the same.

[33] *Progress and Its Problems*, p. 129.

[34] Actually, there is probably room for a *three-fold* distinction. An action or decision may proceed from beliefs and values of the agent and thus be deemed rational in the *personal* sense. If these beliefs and values are the dominant beliefs and values of the time and place, then the action may also be deemed rational in the *social* sense. And if the beliefs and values are themselves reasonable ones, then the action or decision may be deemed rational in the *strong* sense. Galileo's rejection of scriptural evidence against Copernicanism was personally rational, socially irrational, and rational in the strong sense. The Azande tribesman who engages in magical practices may act rationally both in the personal and social sense, though perhaps not in the strong sense. When Laudan talks of the rationality and irrationality of scientific decisions, he has only the social sense of 'rational' in mind.

[35] Laudan claims, for example, that standards of experimental precision have changed. But the undoubted increase in experimental precision need not betoken a change in standards. The standard may always have been that an experiment should be precise enough for the problem in hand ('Sufficient unto the day is the experimental precision thereof'). The increase in precision might then be accounted for by pointing to the more precise experimental problems which are posed by the development of precise theories. (I for one would want to argue against the view that experimental precision

is a virtue in itself, but for the view that theoretical precision is an intrinsic virtue.) For another of Laudan's examples of changing standards which I find problematic, see my 'Problems with Progress', pp. 454–456.

Some standards have undoubtedly changed. Early experimenters did not describe their experiments in such a way that others might reproduce them, or distinguish measured quantities from theoretically calculated ones. On both these counts, standards have changed. And there are some rather obvious arguments to show that the changes are for the better.

Finally, it cannot be denied that there have been great changes, consequent upon changes in theory, in the 'kinds of things which count as empirical problems, the sorts of objections that are recognised as conceptual problems, the criteria of intelligibility, . . . the importance or weight attached to problems' (Laudan, *Progress and Its Problems*, p. 131). But such changes are not the same as changes in methodological *standards*.

36 'History of Science and Its Rational Reconstructions', pp. 117 and 118 (*Philosophical Papers*, I, 132 and 133). Lakatos occasionally suggests that a methodology should seek to explain, not any judgement of the form 'At time *t* theory *A* was better than theory *B*', but rather any such judgement made by the 'scientific élite' ('History of Science and Its Rational Reconstructions', p. 121; *Philosophical Papers*, I, 137). But this appeal to 'élite scientists' raises as many problems as it solves: How are the 'élite scientists' to be identified? Why should their judgements never be mistaken? What if they disagree?

37 McMullin, despite his earlier remarks about 'methodology' needing to be supplemented by 'metamethodology' (see above, Note 28), defends an even broader position. He claims that the 'testimony of history' must be invoked to settle *any* dispute concerning scientific values: 'Our [philosophical] intuitions are simply not secure enough in this domain. They have to be instructed, and supported, and challenged by the testimony of history. Disagreement [over 'strategies of theory-assessment'] is common, and there is no alternative to invoking the historical record. . . . Where intuitions diverge, there is no option but to turn to the historical record, not only to discover what scientists in fact *have* done, but also to sharpen the philosopher's own intuitions . . . . ' ('History and Philosophy of Science: A Marriage of Convenience', pp. 595, 599, 600.)

Now I am all for 'sharpening intuitions', if this means being enabled to support them with arguments. And if the 'historical record' suggests some arguments, or even contains philosophical arguments proposed by scientists, well and good. But how can mere appeal to historical facts settle a dispute about evaluative questions? Consider McMullin's own example, the dispute over whether predicted facts provide better evidence for a theory than previously known facts. The historical record will reveal scientists both explaining known facts and predicting new ones, which will not settle the issue. The historical record may reveal scientists themselves disagreeing on the matter, which will not settle the issue either. What if the historical record reveals that scientists (or a great majority of them) share the philosophical intuitions of one party to the dispute? That may well give the other party pause. But it cannot settle the philosophical issue (unless, of course, we assume that scientific might makes philosophical right). Only philosophical *argument* can settle an issue such as this.

Incidentally, the methodological principle 'Evidence cannot genuinely support a theory unless it is discovered after the theory is proposed' requires for its application historical investigation of the time-order of theory and evidence. But this does not mean that the principle itself can be justified by historical investigation into the behaviour of scientists. (Compare McMullin, 'Philosophy of Science and Its Rational Reconstructions', p. 241.)

[38] This seems to be Noretta Koertge's complaint about Lakatos's extended view: see her 'Rational Reconstructions', pp. 366–367. It was also voiced, prophetically, by Paul Feyerabend: see his 'How To Be a Good Empiricist', p. 18, note. Feyerabend has continued to voice such complaints against the 'historical methodologists', culminating in his question, 'What's so great about science?'. I think he is right to voice the complaints and ask the question, but wrong to assume (as he apparently does) that the answer must be, 'Nothing much'. I have not, of course, argued that point here: for some arguments, see my 'Theorie, Erfahrung und wissenschaftlicher Fortschritt'. Not everybody complains, of course. Toulmin is bemused by Lakatos's vehement opposition to his own view, and seeks to welcome him into the Wittgensteinian fold (see his 'History, Praxis and the "Third World" ').

[39] 'Thomas Young and Newtonian Optics', p. 164.

[40] *Ibid.*, p. 168.

[41] One might say that one optimal methodology, on Worrall's view, is not 'Anything scientists do is rational' but rather 'Nothing scientists do is rational'. On this methodology, anything which history reveals as having influenced a past scientific decision will be an 'external factor'. And historical research will always 'confirm' the Feyerabendian view that 'Anything goes'.

Worrall may object that we may not find historical evidence for either 'internal' or 'external' factors, so that it will not be as easy to 'confirm' a methodology as I here suppose. But where historical evidence exists, it cannot fail to confirm any methodology. And where it is lacking, through gaps in the historical record, why should this count against any methodological view?

[42] In practice, this is what Worrall does. All the 'external explanations' which he considers in his fascinating historical case-study invoke 'external factors' in this traditional sense.

[43] 'History of Science and Its Rational Reconstructions', p. 118.

[44] For the general procedure, see Lakatos, 'History of Science and Its Rational Reconstructions', p. 107 (*Philosophical Papers*, I, 120). For the examples, see 'Falsification and the Methodology of Scientific Research Programmes', pp. 138 and 146 (*Philosophical Papers*, I, 53 and 60–61); and 'History of Science and Its Rational Reconstructions', p. 106–107 (*Philosophical Papers*, I, 119).

[45] See Kuhn, 'Notes on Lakatos', p. 143 (also 'Reflections on My Critics', p. 256, Note 1); Holton, *The Scientific Imagination*, p. 106; McMullin, 'History and Philosophy of Science: A Marriage of Convenience', p. 598; Laudan, *Progress and Its Problems*, pp. 168 and 170.

[46] Of course, Lakatos may have been wrong to attribute to Bohr in 1913 ideas which had electron-spin as a natural consequence or extension. That he could well be wrong is indicated by the very fact that Bohr accepted electron-spin only in 1926. But scientists never draw all the consequences of their own ideas. And it is one thing to object to Lakatos's history, another to object to his colourful way of presenting it.

[47] 'Falsification and the Methodology of Scientific Research Programmes', pp. 179–180 (*Philosophical Papers*, I, 92); 'History of Science and Its Rational Reconstructions', p. 106 and Note 60 (*Philosophical Papers*, I, 118 and 119, Note 1).

[48] Here, and in what follows, I am indebted to Gregory Currie's 'The Role of Normative

Assumptions in Historical Explanations'.

49   Suppose that two rival methodologies M and M^* happen to agree that theory A was preferable to theory B at time t. And suppose A was preferred to B at time t. Worrall will say that both M and M^* are 'confirmed'. I would say that so far neither is 'confirmed'. M would be 'confirmed' and M^* 'refuted' if historical investigation revealed that scientists applied methodology M, and not methodology M^*, in reaching their decision. Both M and M^* would be 'refuted' if it turned out that the decision was based on yet a third methodology. (I put scare-quotes around 'confirm' and 'refute' here because it is the historical claim which Worrall associates with each methodology which gets confirmed or refuted, not the methodology itself.)

50   Perhaps I might add here that in my own essay in 'philosophically inspired history of science' I tried to show (through no doubt I did not try hard enough) that the preference for Lavoisier's oxygen theory of combustion over the phlogiston theory was based on an acceptance of some of Lakatos's normative principles. I did this in a joking way, saying that by 1777 Lavoisier had obviously read Feyerabend and that by 1783 he had read Lakatos too ('Why did Oxygen Supplant Phlogiston?' pp. 197 and 203). The joke was meant to suggest that those who read Lavoisier's arguments and found them convincing were implicitly adopting methodological norms later articulated by Feyerabend and Lakatos. But as I recall the original discussion of that paper, the joke fell upon stony ground and I was accused of dabbling in 'second-world considerations' (and implausible ones to boot).

   Since I have mentioned that paper, I will take the opportunity to mention an error which it contains (which has been pointed to me by Homer Le Grand and John Elliot). I claim that Lavoisier predicted the discovery of oxygen (*op.cit.*, p. 192). In fact, he did not (though he might have done); the discovery merely fitted nicely into his scheme.

51   For this modified position, see Lakatos, *Philosophical Papers*, I, 191–192. Also Worrall, 'Thomas Young and Newtonian Optics', pp. 164–168.

   Curiously enough, Laudan, despite all his strong words about 'rational reconstructions', falls into a view similar to the Lakatos-Worrall view: see my 'Problems with Progress' and Currie's 'The Role of Normative Assumptions in Historical Explanation'.

52   One can, of course, formulate an historical explanation in such a way that an evaluative premise figures essentially in it. Consider:

(E)   At time t theory A was better than theory B.
       Scientists believed this at t and acted accordingly.
       Therefore, at t scientists preferred theory A to theory B.

But such formulations are doubly misleading. If historical investigation reveals that the conclusion is false, we might be tempted to direct *modus tollens* at the first premise and mistakenly 'refute' a normative judgement by a historical fact. If philosophical discussion convinces us that the first premise is mistaken, we might be tempted to conclude that (E) is mistaken *qua* historical explanation; yet there might be compelling historical evidence that at time t scientists did endorse the first premise! Obviously, the *explanatory* power of (E) is fully captured by:

(E*)   At time t scientists believed that theory A was better than
       theory B and acted accordingly.
       Therefore, at t scientists preferred theory A to theory B.

[53] Again, I owe this point to Gregory Currie: see his 'Popper's Evolutionary Epistemology: A Critique'. In the above, I have taken issue only with the view that Popper's 'second world' is *causally* influenced by his 'third world'. I have not disputed the view that the physical world is causally influenced by the mental, and hence is not 'causally-closed'. But of course, those who seek a physicalist account of mental phenomena will thereby hope to re-establish the causal closure of the physical world.

[54] One glib argument is that before the historian of science can get started he must decide what science is (or who is a scientist), which will involve him in solving the philosophical problem of demarcation, which will involve him in elaborating a whole philosophy of science. (For this argument, see Lakatos, 'History of Science and Its Rational Reconstructions', pp. 120—121.)

The obvious answer to this is that as a matter of historical fact some people have come to call themselves 'scientists' and to distinguish their pursuits from other pursuits. How and when this came about is an interesting historical question in its own right. The historian of science can simply say that his subject-matter has delimited itself. And he can go on to ask why the history of science should not be as 'value-free' as political or economic history. We can ask what happened to the economy or the political institutions of eighteenth-century France without also asking whether it was a good thing. Why cannot we ask what happened to chemistry in eighteenth-century France without also asking whether it was a good thing?

Of course, a philosopher armed with a 'criterion of demarcation' might well argue that some historically-given science is not worthy of the name. And he might be right: was phrenology a science in the nineteenth century? Is scientology a science today? But this is not history any more, but a philosopher laying down the law.

REFERENCES

Albert, H. and Stapf, K. (eds.) (1979) *Theorie und Erfahrung*, Stuttgart: Klett-Cotta Verlag.

Baumrin, B. (ed.) (1963) *Philosophy of Science: The Delaware Seminar Volume II*, New York: Interscience Publishers.

Beck, M. T. (1978) Editorial Statement, *Scientometrics* 1, 3—4.

Briskman, L. B. (1974) 'Toulmin's Evolutionary Epistemology', *Philosophical Quarterly* 24, 160—169.

Brown, J. (1977) 'Moral Theory and the Ought-Can Principle', *Mind* 86, 206—223.

Buck, R. C. and Cohen, R. S. (eds.) (1971) *PSA 1970, Boston Studies in the Philosophy of Science*, Vol. 8, Dordrecht: D. Reidel Publishing Co.

Cohen, R. S., Hooker, C. A., Michalos, A. C., and Van Evra, J. (eds.) (1976) *PSA 1974, Boston Studies in the Philosophy of Science*, Vol. 32, Dordrecht: D. Reidel Publishing Co.

Cohen, R. S., Feyerabend, P. K., and Wartofsky, M. W. (eds.) (1976) *Essays in Memory of Imre Lakatos, Boston Studies in the Philosophy of Science*, Vol. 39, Dordrecht: D. Reidel Publishing Co.

Colodny, R. G. (eds.) (1966) *Mind and Cosmos*, (University of Pittsburgh Series in the Philosophy of Science, Volume III, Pittsburgh: University of Pittsburgh Press.

Currie, G. (1978) 'Popper's Evolutionary Epistemology: A Critique', *Synthese* 37, 413—431.

Currie, G. (1980) 'The Role of Normative Assumptions in Historical Explanation',
  Philosophy of Science 47, 456–473.
Feyerabend, P. K. (1963) 'How To Be a Good Empiricist' in Baumrin (ed.), pp. 3–39.
Feyerabend, P. K. (1975) *Against Method,* London: New Left Books.
Feyerabend, P. K. (1976) 'On the Critique of Scientific Reason', in Howson (ed.), pp.
  309–339.
Giere, R. N. (1973) 'History and Philosophy of Science: Intimate Relationship or
  Marriage of Convenience?', *British Journal for the Philosophy of Science* 24, 282–
  297.
Gilbert, G. N. (1978) 'Measuring the Growth of Science: A Review of Indicators of
  Scientific Growth', *Scientometrics* 1, 9–34.
Holton, G. (1978) *The Scientific Imagination,* London: Cambridge University Press.
Howson, C. (ed.) (1976) *Method and Appraisal in the Physical Sciences,* London:
  Cambridge University Press.
Jarvie, I. C. (1976) 'Toulmin and the Rationality of Science', in R. S. Cohen *et al.*
  (eds.), *Essays in Memory of Imre Lakatos,* pp. 311–333.
Jevons, F. R. (1973) *Science Observed,* London: George Allen and Unwin Ltd.
Koertge, N. (1976) 'Rational Reconstructions', in R. S. Cohen *et al.* (eds.), *Essays in
  Memory of Imre Lakatos,* pp. 359–369.
Kuhn, T. S. (1962) *The Structure of Scientific Revolutions*, Chicago: University of
  Chicago Press (Second Edition, with 'Postcript – 1969', 1970).
Kuhn, T. S. (1970) 'Logic of Discovery or Psychology of Research?', in Lakatos and
  Musgrave (eds.), pp. 1–23.
Kuhn, T. S. (1970) 'Reflections on My Critics', in Lakatos and Musgrave (eds.), pp.
  231–278.
Kuhn, T. S. (1971) 'Notes on Lakatos', in Buck and Cohen (eds.), pp. 137–147.
Kuhn, T. S. (1977) *The Essential Tension,* Chicago: University of Chicago Press.
Lakatos, I. (1970) 'Falsification and the Methodology of Scientific Research Pro-
  grammes', in Lakatos and Musgrave (eds.), pp. 91–195 (and in his *Philosophical
  Papers*, I, 8–101).
Lakatos, I. (1971) 'History of Science and Its Rational Reconstructions' in Buck and
  Cohen (eds.), pp. 91–135 (and in his *Philosophical Papers* I, 102–138).
Lakatos, I. (1978) 'Understanding Toulmin', in his *Philosophical Papers,* II, 224–243.
Lakatos, I. (1978) *Philosophical Papers,* Volumes I and II, edited by J. Worrall and G.
  Currie, London: Cambridge University Press.
Lakatos, I. and Musgrave, A. (eds.) (1970) *Criticism and the Growth of Knowledge,*
  London: Cambridge University Press.
Laudan, L. (1977) *Progress and Its Problems,* London: Routledge and Kegan Paul Ltd.
McMullin, E. (1976) 'History and Philosophy of Science: A Marriage of Convenience', in
  R. S. Cohen *et al.* (eds.) *PSA 1974,* pp. 585–601.
McMullin, E. (1978) 'Philosophy of Science and Its Rational Reconstructions', in
  Radnitzky and Anderson (eds.), pp. 221–252.
Musgrave, A. E. (1972) 'George Boole and Psychologism', *Scientia* 107, 593–608.
Musgrave A. E. (1976) 'Why Did Oxygen Supplant Phlogiston?', in Howson (ed.),
  pp. 181–209.
Musgrave, A. E. (1979) 'Problems with Progress', *Synthese* 42, 443–464.

Musgrave, A. E. (1979) 'Theorie, Erfahrung und wissenschaftlicher Fortschritt', in Albert and Stapf (eds.), pp. 21–53.

Narin, F. (1978) 'Objectivity versus Relevance in Studies of Scientific Advance', *Scientometrics* 1, 35–41.

Passmore, J. (1983) 'Why Philosophy of Science?', in this volume pp. 5–29.

Popper, K. R. (1963) *Conjectures and Refutations*. London: Routledge and Kegan Paul.

Radnitzky, G. and Anderson G. (eds.) (1978) *Progress and Rationality in Science, Boston Studies in the Philosophy of Science, Vol. 58,* Dordrecht: D. Reidel Publishing Co.

Roszak, T. (1972) *Where the Wasteland Ends,* New York: Doubleday.

Roszak, T. (1973) 'Some Thoughts on the Other Side of This Life', *The New York Times,* 12 April.

Shapere, D. (1966) 'Meaning and Scientific Change', in R. G. Colodny (ed.), pp. 41–85.

Toulmin, S. E. (1958) *The Uses of Argument,* London: Cambridge University Press.

Toulmin, S. E. (1976) 'History, Praxis, and the "Third World"', in R. S. Cohen *et al.* (eds.), *Essays in Memory of Imre Lakatos, pp.* 655–675.

Watkins, J. W. N. (1978) 'Corroboration and the Problem of Content-Comparison', in Radnitzky and Anderson (eds.), pp. 339–378.

Williams, L. P. (1975) 'Should Philosophers of Science Be Allowed to Write History?', *British Journal for the Philosophy of Science* 26, 241–253.

Worrall, J. (1976) 'Thomas Young and Newtonian Optics', in Howson (ed.), pp. 107–179.

BRYAN GANDEVIA

NO HISTORY WITHOUT HEALTH

I had misgivings in agreeing to address this meeting because of some doubt that I could contribute on the same plane as other contributors. Having now heard the earlier papers I am convinced that my observations must necessarily be at a more mundane level. In other company I might well venture into the philosophy of medicine and medical history, a philosophy which is inherently broader in its concepts and foundations than the philosophy of science and scientific history, even though there is considerable overlap.[1] To understand the scientific and humanitarian concepts of medicine, as well as the approaches to research and practice in medicine in any society or culture, demands an understanding of contemporary philosophy. At the same time, perhaps even more than in the history of science, it is possible to adopt a reasonably pragmatic approach. Dr. Crookshank in a slightly tongue-in-cheek introduction to Cumston's *History of Medicine* wrote:

... so, a dose of castor oil acts with equal efficiency whether given to expel a demon, to calm the vital spirits, to assuage the Archaeus, to evacuate morbific humours, to eliminate toxins, to restore endocrine balance, or to reduce blood pressure.[2]

A modern sceptic might add also 'to produce iatrogenic disease'.

For present purposes my definition of Medicine must be a very broad one which embraces not only problems of disease and its management but also the concept of 'positive health', a semantically indefensible term with enormous bureaucratic appeal, especially since it has received the imprimatur of the World Health Organization. Given a holistic definition, the history of medicine permeates all aspects of history. In Australian art, for example, notably in the work of McCubbin, Streeton and Drysdale, but also in that of many others, there are motifs which require some understanding of the medical history of this country; or, putting this round another way, the artist in his picture is offering a comment on the medical (or social) problems of his day. Much Australian literature and poetry, especially when deliberately orientated as social comment, contain allusions which can be fully appreciated only with some knowledge of background hazards to health and welfare: as, for example, the poetical works of Henry Lawson and, on a rather different level, those of Bernard O'Dowd. Except occasionally in the latter,

R. W. Home (ed.), Science Under Scrutiny, 81–98.
© *1983 by D. Reidel Publishing Company.*

one rarely sees the refinement of medical imagery, simile and metaphor to be found in Shakespeare's work, perhaps most strikingly in *Troilus and Cressida*, the study of which first directed my attention to the ubiquity of medical history.[3] In Ben Jonson's plays medical matters are often dealt with in more explicit fashion, as also in Molière and Shaw.

THE HISTORIOGRAPHIC ROLE OF THE HISTORY OF MEDICINE

These introductory remarks serve only to indicate that I do not see the history of medicine as any narrow discipline solely confined to tracing progress in its science and technology; any effective teaching programme must take cognisance of the cultural and social relationships of medicine. Without wishing to open the philosophic floodgates, I believe one may see history simply as the story of man's endeavour to adapt himself to his environment. If one accepts this observation, it becomes apparent that the indices of adaptation must ultimately be primarily medical, again using that term in a broad sense to include indices of health as well as of disease, in populations or societies as well as in individuals. To take an elementary illustration of what I mean, let us consider the heights of the convict children transported to Australia *circa* 1820 — 1840. Boys aged between 13 and 18 years were approximately 6 — 9 cm shorter than other recorded series of children in Britain of about that period, and even shorter than grossly underprivileged children in various institutions.[4] The short stature reflects their social and physical environment, perhaps particularly its nutritional component. Although one may wish to make allowance for artistic licence in presenting convicts as a lower kind of animal, it is a fact that scenes showing convicts almost invariably show them as of lesser stature than their guards and with singularly ugly physical characteristics. The facies of a few of the convicts recorded in contemporary sketches suggests that of a child born to an alcoholic mother (an idea I owe to Dr. Gillian Turner); it will be interesting to see if this concept can derive any support from a more sophisticated analysis of the recorded physical characteristics of convict children than has hitherto been made.[5] Finally, as a reflection of the influence of environment, all the heights of first generation Australians which we have been able to find lie, with one exception, more than two standard deviations above the regression line relating height to age in the convict children. This is not the place to examine the reasons for this rapid change but let us also note that these physical measurements are fully in accord with contemporary descriptions of the physique of the young 'Cornstalks'.

One may represent this interacting or 'feedback' conception of the role of the history of medicine in the following simplistic diagram:

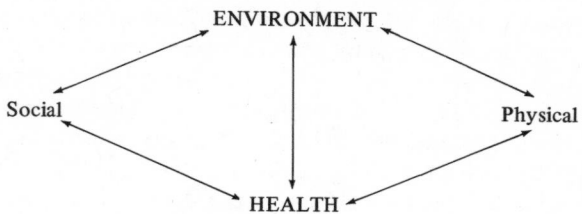

We may list the social environmental factors requiring consideration as follows: population structure; organisation and government (control, discipline and morale); education and literacy; concepts of disease, health and death (religious, philosophical and scientific contributions); economic structure and status; previous experience of disease(s); and technological development (notably insofar as it permits environmental control). In the physical environment, obviously relevant factors are terrain, climate, food supply, water supply, housing, isolation and communication. Doubtless there are other factors. In this context 'factor' does not have the precise significance which it has in mathematics; when the word is used in the social sciences it is almost always a manifestation of ignorance. For each factor, one requires an *index*, which may be either quantitative or qualitative (height, facies, literary descriptions were three indices used in the foregoing discussion of the convict children).

I am aware of the arguments concerning the role of statistics and quantitative estimates[6] in social history, but most of the arguments against them are facile provided the statistics are used with exactly the sáme precautions and reservations as they are in medicine or science. To regard statistics as the essence of history is as naive as it is to ignore quantitative data when these are available. Quite apart from the principle of quantification or measurement, there is the probability (but no more) that measurements will be more objective than mere description or qualitative observation. It seems to me that to argue for a dichotomy is about as futile as the argument over history as a science or an art; both types of evidence are admissible just as both kinds of approach to history are inseparable.

The selection of indices to examine social and environmental factors in history is a specialised task, and ideally should be done by specialists from the several disciplines, implicit in my list of factors, which are involved on either the physical or the social side. It is likely that a geographer will define indices of climate and terrain better than a political scientist or an economist, and a specialist from demography or medicine may provide the best indices of health; although when the concept of health and disease is extended to populations one might sometimes wish for the indices to be defined by a sociologist or a social psychologist. I came across the latter problem in attempting to find indices which would reflect the state of morale in the first settlement; crime, marriage and illegitimacy rates are among many which may be used.[7] No matter from what specialty the historian is drawn, it is essential that he define his terms and his indices so that they may be applied in a comparable manner to other situations, other periods and other places. This should be intuitive in those from strictly scientific disciplines but it is often overlooked by those from the more social or 'arts' disciplines.[8]

There is of course a dimension of time in history, and this could be represented in a three-dimensional version of my 'model' above. Historians from some disciplines handle the dimension of time much better than others; time is an integral part of disciplines such as epidemiology, demography and geography but is less so in the social sciences, where orientation and concern are often localised to the present or to a limited period. History as written by historians from the latter disciplines often reveals the mistake of 'explaining' the past in the light, or rather perhaps the penumbra, of modern concepts and hypotheses.[9] This is well seen where value judgements are applied to such subjects as attitudes to the charitable works and welfare institutions of, say, the nineteenth century; these almost inevitably imply that today's concepts of charity and social welfare and their practice are 'correct', whereas hopefully today's concepts are not immutable but will reach a more mature level of development in decades to come.[10]

EXAMPLES

My intention now is to give a variety of examples illustrating the foregoing from my studies of the history of medicine and health in Australia. My approach to these examples will be superficial, as it is not my purpose to teach history or even a concept of historiography; for those concerned with teaching history I am anxious to offer a way of looking at the subject which is often able to provoke a reappraisal of accepted views.

The mortality of convict transportation shows interesting fluctuations over the best part of a century. The first fleet arrived after a long voyage but with what was, for the period, an acceptable mortality rate. This could not be said for the second fleet nor for many later transports. Figure 1 shows that after certain administrative measures were taken in 1815 there was a steady decline in mortality. However, this improvement was followed by a rise in mortality, associated with an increase in the number of convicts transported, up to about the mid-1830s. Yet the correlation of mortality with increasing numbers transported is less impressive than the correlation of the former with an arbitrary index of overcrowding, namely, the tonnage of the ships divided by the number of convicts transported.[11] There are always difficulties in the interpretation of correlations but there are other reasons and other descriptive evidence which suggest that overcrowding contributed to a rising mortality, and the graphical demonstration may be viewed as doing no more than conforming to this hypothesis. There were other factors involved; one was the progressive reduction in the time taken for the journey.

The mortality over the first five years or so of settlement at Sydney offers a fascinating field for study (Figure 2).[12] There was initially a small epidemic

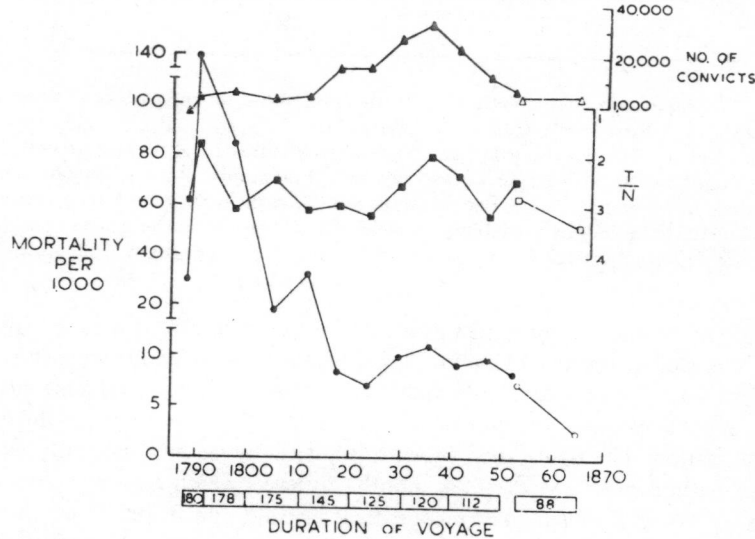

Fig. 1. Mortality (lower line) on convict transports, 1788–1865. The upper line shows the number of convicts embarked, and the intermediate line the 'space factor' T/N (ship tonnage divided by number of convicts). The average duration of the voyages in days is shown in the lower panel. Closed circles relate to voyages to Sydney or Hobart and open circles to Perth voyages. (Reproduced by permission from *Med. J. Australia*, 1967, 2, 941.)

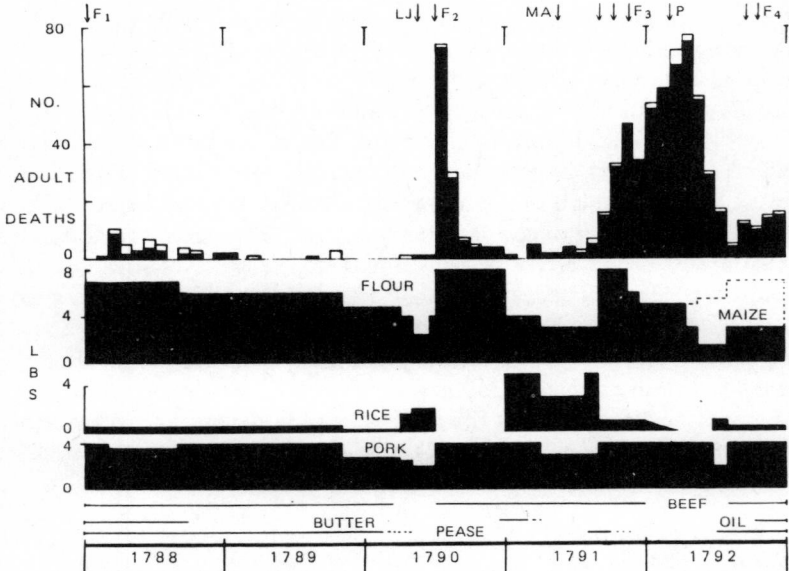

Fig. 2. Explanation of graphs and symbols from above down. Arrows indicate ship
arrivals – F_1, First Fleet; L.J., *Lady Juliana*; M.A., *Mary Ann*; F_2, Second Fleet; F_3,
Third Fleet; P., *Pitt*; F_4, Fourth Fleet. Vertical bars indicate end of year on horizontal
scale. Adult deaths, number on vertical axis shown by month. Rations are indicated on
the three lower graphs. Flour, rice and pork are shown to scale in lbs. Horizontal lines
below indicate periods of availability or otherwise of beef (issued in place of part or all
of pork ration), butter, oil and pease. (Reproduced by permission from *Aust. N.Z.
J. Med.*, 1974, 4, 111.)

of dysentery and scurvy, much as would be expected after a long sea voyage
and the establishment of a camp. This epidemic occasioned little concern and
it soon settled down. Mortality for the next two years was negligible despite
the fact that towards the end of this period the settlement was on the verge
of starvation. The figure then shows a dramatic increase in mortality with a
rapid decline over two or three months to levels which are again very low.
This epidemic involved only second fleet arrivals, and the shape of the epidemic curve (a dramatic upsweep with a subsequent steady decline) strongly
suggests that the epidemic commenced prior to the arrival of the fleet; in fact,
other evidence indicates that this is so. The deaths in this case were probably
due to multiple diseases but primarily to the introduction of these diseases
amongst an overcrowded, malnourished and ill-treated population transported
in the most unhygienic circumstances.

Despite another period of inadequate rations, the mortality rate then returns to the same level as before. Following the arrival of the third fleet, there is a remarkable epidemic extending over a period of nearly a year. Detailed analysis has indicated that almost all those who died were drawn from those who arrived on the third fleet. There are no medical descriptions of the illness or illnesses from which the people suffered, although there are some lay observations concerning the manner of death. Again the mortality is independent of the ration level, so that lack of food or malnutrition was not the sole cause of death although both may have contributed to deaths from other causes. I shall not here discuss the nature of this illness, which produced a mortality as high as in the worst period of the great plague of London. Suffice it to say that it had a tremendous impact on the function and morale of the colony.

The mortality experience of convicts arriving on the early fleets is summarised in Figure 3. This shows the survival rates, and the outstanding feature is the extraordinarily high survival rate for the first fleet, a survival rate unlikely to have been reached had these convicts remained in Britain. The

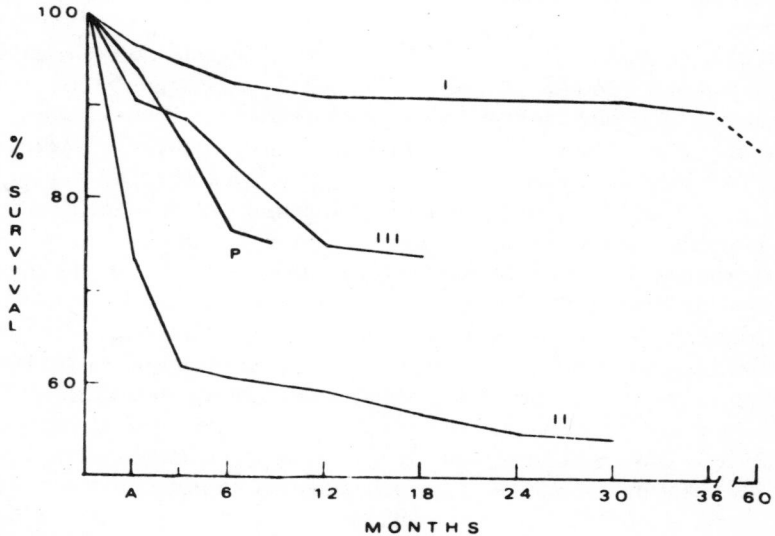

Fig. 3. Survival rates for the First, Second and Third Fleets and the *Pitt*. 100% represents the number embarked in Britain; A, percentage surviving on arrival at Port Jackson, Sydney; thereafter, the lines show the percentage survival at six-monthly intervals from the time of arrival. (Reproduced by permission from *Aust. N.Z. J. Med.*, 1974, 4 111.)

horrible mortality of the second fleet is clearly indicated and it will be noted that this group never really adapted to its environment, in that after arrival the survival rate continues to fall at a greater rate than that for the first fleet. The third fleet mortality is obviously high but it is not possible to follow those involved for very long after their arrival in the colony.

No history of the first settlement should ignore such a remarkable mortality (and morbidity) record, although one or two histories, including Manning Clark's, do. An explanation must be found in the social and physical environment of the settlement, and the historian should surely provide it. The problem is most challenging in relation to the third epidemic, and this must be contrasted with the remarkable adaptation to the environment shown by members of the first fleet. I believe that the answer lies simply in the sound leadership of Phillip and the morale which he managed to engender, despite adversity, amongst the first and to a lesser extent the surviving second fleet convicts. He had little opportunity to do this with the third fleet, and for the period following his departure all indices of morale (that is to say, of the psychological welfare of the population) show an abrupt decline. This reflects the influence of the poor leadership which characterised the interregnum.

A notable feature of the first few years of settlement was the relatively low infant and child mortality, leading to a relative increase of children in the population. This gave contemporaries the impression of enhanced fertility, an impression for the most part not borne out by careful study of the figures.[13] Improved survival of infants is a well-recognised index of adaptation to the environment, and the experience must have seemed strange to mothers from the industrial slums of England. There is evidence that this survival rate was better amongst the relatively stable 'families' (whether blessed by the church or not) than amongst the children of more transient liaisons.

Deaths from trauma showed a pattern which might be expected in a penal settlement in an alien land — executions, death at the hands of the aborigines or due to starvation in the bush dominated the picture, while suicide, using the crude and very traumatic means available to the convicts, remained surprisingly low. A high mortality rate from accidents at all ages remained a fairly characteristic feature of Australian mortality throughout most of the nineteenth century.[14] Patterns of malingering, as well as its prevalence, were influenced by the harshness of the penal (social) environment.[15] During the worst periods of the Norfolk Island settlement, only the most severe forms of self-mutilation stood any reasonable chance of being accepted by authority, and hence of offering a somewhat easier way of life. Common methods employed to this end were putting lime or a toxic plant juice in the eyes to produce acute inflammation and blindness (which was not always temporary), the production of self-inflicted traumatic ulcers (facilitated by the poor nutritional state and the use of leg irons), and the ingestion of toxic materials, notably plant glycosides, to produce mental derangement. Assigned servants in remote areas did not resort to chopping off toes when their objective was a long walk to a centre of civilisation; to an employer who was not medically qualified it was obviously better to complain of palpitations, weakness or other internal complaints.

Australia is one of the few countries in the world in which malingering has become so socially acceptable as to be built into many industrial awards, whereby one is entitled to five days' leave per year without a medical certificate. The right to a 'sickie' takes away any originality in terms of history or physical signs, since it is not necessary to simulate anything.

One of the reasons given for the withdrawal of Captain Maconochie from Norfolk Island was his alleged failure to obtain more productive work out of the convicts under his command. A study of the morbidity and mortality rates from dysentery makes it clear that he had a work-force which was capable of very little work, and certainly incapable of reaching the targets set by remote administrators. Even more interesting is the fact that recurrent epidemics of severe dysentery swept through the 'new hands', the convicts specially sent out from England direct to Norfolk Island to participate in a trial of Maconochie's new system of prison discipline, whereas over the period of his administration dysentery was not a significant problem amongst the secondarily convicted 'old hands' from New South Wales. The reasons for this are complex but at least the relative immunity of the old hands (considering that they were treated and lived no differently from the new hands) accords well with the clean and smart appearance which excited the comment of other observers besides Governor Gipps. It is of course apparent that, in the circumstances of the Norfolk Island settlement, neither flies, food, water supply, milk nor similar vectors could have been responsible since everyone was equally exposed to any such hazard. Differences in personal hygiene seem the only likely contributory factor.[16]

I mentioned earlier some of the physical characteristics of the convicts but it is especially interesting to consider the phenomenon of tattooing.[17] As this 'disease' has a negligible mortality and zero recovery rate, the prevalence will tend to rise in any given population if the practice is continued. Detailed records of the convict tattoos are extant because of their value in identification. The high prevalence of certain kinds of tattoos, notably dots and rings on the fingers, reflects clearly the inferior social status of the convicts; tattoos of this kind are commonly a psychopathic, delinquent or criminal characteristic. Quite apart from indicating the lowly background of many of the convicts, these types of tattoo can be shown to be associated with a disproportionate frequency of the large 'fines' awarded by Maconochie for bad behaviour. Another physical characteristic, pock-marks, implied an origin overseas, since smallpox was never of major importance to the European population in Australia; in association with short stature, tattoos, or an unpleasant visage, pock marks were stigmata implying a convict origin.[17]

The pattern of disease in any community is to a large extent determined by the age distribution of its population. Thus, Australia's aging population ensures that 1938 — 1988 is the era of the chronic diseases. Perhaps the most outstanding feature of nineteenth-century Australian demography was the high birth rate in the immediate post-gold-rush era, followed in due course by a relatively high proportion of children and young adults in the population. One phenomenon associated with these changes was an epidemic of tuberculosis, affecting particularly the young, and reaching its peak in the 1880s. For the tubercle bacillus, as for other agents of infectious diseases, the population was a relatively virgin (non-immune) field. Diseases like rubella and measles were also able to find susceptible adults, especially as the relatively small and scattered population did not at first permit these disorders to establish themselves as endemic. It soon became possible to recognise congenital deafness as an epidemic disease, and retrospectively it is possible to identify mortality from measles in advanced age. There are many other social and environmental factors involved in the epidemiology of the infectious diseases, and both physical and social environmental factors are also involved in the behaviour of the gastrointestinal diseases. Broadly speaking, the latter showed a progressive decline over the period 1800 — 1920, for the most part before many therapeutic or prophylactic procedures were available. For the several different disorders involved (dysentery, typhoid fever and infantile gastroenteritis), as well as for infant mortality itself, the shape of the mortality curves was similar, although the decline began at different periods for different diseases. The definitive work has yet to be done on the morbidity and mortality of these diseases, in relation to the social and physical environmental factors which may have played a part in their declining importance. Closer analysis of statistical data, and especially comparative studies from different centres of population, may lead to a more precise understanding of the relative importance of a variety of contributory influences. For example, it seems most unlikely that improvements in personal and family hygiene could occur until there was some relevant teaching in the schools and a measure of literacy amongst adults from the poorer classes; by the last decade or so of the nineteenth century these prerequisites were beginning to be achieved. Equally, the development of sanitation cannot explain different rates of decline of morbidity and mortality for the several gastrointestinal disorders.[18]

SPECIALIST HISTORY AND HISTORIANS

The specialist historian may contribute to the main stream of historiography and history in several ways.[19] Firstly, he may provide the technological history of his specialty. This is best done by a specialist trained in the field of enquiry, though this does not preclude significant contributions, particularly of a synthetic or philosophical kind, from people trained in other disciplines. It was not long ago that this kind of scientific (or medical) history was decried by social historians as little more than antiquarianism. In some respects it is, but this in no way diminishes the need for it. In its narrowest form, the history of a science or technology, or of medicine (narrowly defined), serves at least to illustrate the development, both technical and philosophical, of the subject. The student of that specialty may by this means gain some perspective on the present state of his profession. The social or general historian, or perhaps more precisely any interdisciplinary historian, may well seek to place this kind of history in a broader perspective, and relate it, for example, to social, political, economic or cultural changes in the environment; or he may seek to relate a technological change to scientific philosophy or evolution. It is essential that such an interdisciplinary historian be provided with first-class technological history to use for his own purposes. This does not absolve the general historian from exercising a critical faculty, but it should help him avoid the florid technical errors or oversights which may otherwise occur.

The second role of the specialist historian and his history is to provide a technical appreciation of changing concepts and attitudes without the specialty, whether these are socially, scientifically or technically based. I can illustrate this best from the field of industrial hygiene and occupational disorders. It is obvious that medical, economic, geographic and legal historians may take very different views of social attitudes to the work situation, as reflected, for example, in workers' compensation procedures, industrial hygiene legislation, or safety regulations. In evaluating the historical role of this kind of legislation, consideration must be given to medical knowledge before, during and after its introduction. The medical situation in the gold-mining industry at the turn of the last century was such as to justify fully the introduction of legislation concerned with prevention of disease and with compensation. On the other hand, the medical situation has now changed, and the worker receiving compensation today for coal pneumoconiosis or silicosis lives longer on average than other males, at least in New South Wales. There are important medical components to the history of 'dirt' or 'danger' money, just as there are legal and hygienic components, and these must all be taken into account if we are to reach a full understanding of the social history of these concepts. Again reverting to the nineteenth century, there was a widespread belief that milk inhibited the development of certain dust diseases, and as a result a daily milk ration became written into some industrial awards. Medical knowledge now indicates that milk has no place in the prevention of pneumoconioses, but I have seen a strike when the milk ration was not delivered on time, despite the fact that on the days when it was, almost all the potential consumers disposed of it by tipping it into the sea. To understand the historical evolution of that strike, it is necessary to understand the medical history, and also to appreciate that from the point of view of preventing disease, it did not matter how the milk was disposed of.

There is another side to this question of the influence of a technology (in which for the moment I include medicine) on the development of social history, in that very often the absence of legislation to control hazards or problems is related to a lack of effective medical knowledge, or a lack of appropriate technology to monitor or control the problem. This must be borne in mind in any historical analysis of social legislation or attitudes in any particular period and community.

The third contribution which the specialist historian may make to general history is to help maintain technical historiographic standards amongst general historians, for his discipline is no less rigorous than that of the professional or trained historian. However, it has become increasingly evident that the epidemiologist in particular, whose scientific discipline includes the dimension of time and who is necessarily historically disciplined, has a special contribution to make to general history and particularly to social history. Historians are beginning to use techniques which are essentially those of the epidemiologist. The epidemiologist has developed a highly sophisticated methodology and has considerable experience and an advanced understanding of sampling methods and statistical procedures. He is aware of the potential for error inherent in the design of his study, and is conscious of the degree of probability which he may attach to generalisations based on data of known validity or limitations. He is familar with the risks of reaching wrong conclusions through faulty procedures. All these questions are just as relevant to historians using epidemiological methods to solve their problems as they are to authorities concerned with the public health. Indices of any of the social and physical environmental 'factors' listed earlier must be rigidly defined, and they must be measured or assessed in equally rigid and well-defined ways, and as objectively as possible. Editors of scientific journals are more scrupulous in regard to the use of key words than are the editors of historical journals. Social historians, especially since the rise of interest in oral history, are using techniques which accord with epidemiological concepts of cohort studies, cross-sectional surveys and even prospective surveys. There is a vast amount of factual information available on these techniques and even the historian who does not use them needs this knowledge in order to evaluate critically the work of those who do. I was interested to learn that in two geography departments in Australian universities there are courses in what might be termed historical epidemiology. I suggest that courses in epidemiology, preferably historical epidemiology, should be included in history courses, with the emphasis lying on methodology. To take a simple example of the importance of this, in my first figure there are three epidemic curves. The third is characterised by a tolerably normal distribution of mortality, that is to say, a more or less steady increase towards a peak and then a more or less equally steady decline. The second fleet epidemic curve, by contrast, shows a very sudden upswing, but if weeks instead of months had been used as the base line it would have shown a steady smooth decline; that is to say, it resembles only the right half of the 'normal' epidemic curve. The explanation of these epidemics and their course must therefore be entirely different, even if the same diseases happened to be involved (although usually one would expect them to be different). To deal effectively with the history of this quinquennium it is essential to appreciate the different types of epidemic curve, and to understand the reasons for the differences between them.

A fourth function of specialist history, in some ways the most important, is also one which I have examined previously on a number of occasions. However, I am not sure that the importance of the principle involved is fully appreciated. Whatever indices of health or disease we choose to study, the conclusions drawn from them must be in accord with what is known of the physical and social environment; it is impossible that the data from the social and physical environment should ultimately be at variance with the health indices, since all come from the same community. It thus becomes possible not simply to formulate a hypothesis (in scientific terms) or, in historical jargon, offer an explanation, but also to test its truth or validity. If it it suggested that the decline in various diseases is attributable to improvements in water supply and sanitation, then it is reasonable to ask whether there was any temporal correlation between these socio-environmental developments and the mortality rates from these diseases. When we find that such correlations were far from close, some other factors must be postulated, for example, rising literacy and thus the possibility of mass education in personal hygiene. If I offer medical data, and an interpretation, to suggest that much of the mortality of third fleet arrivals was due to loss of the will to live, then there should be many other indices, not medical but social, of this social malaise. I would expect this important function of specialist history to be equally applicable to the social and physical environmental relationships of much scientific and technological history. It is most unlikely that the stump-jump plough would be invented in a nomad society where the climate and terrain did not encourage the growth of crops. By contrast, such an invention might tip the balance just sufficiently in favour of a community to ensure its survival in adverse circumstances.

MEDICAL HISTORY AND MEDICAL EDUCATION

As in many other fields of knowledge and endeavour, the specialist historian runs the risk of oversimplification whenever he transgresses interdisciplinary bounds, whilst the generalist, although he may well develop broader insight and perspective, runs an even bigger risk of a similar kind. These considerations pose problems not so much for research in the history and philosophy of science, medicine and technology, but for learning and teaching them.[20] Insofar as medicine is concerned, the practical problems of teaching the history of medicine have been greatly augmented by shortened and crowded courses and, in some universities, by a sort of 'integrated' teaching (so-called), a regression to the ancient Egyptian concept which permitted the classification of surgical diseases in a regional fashion from the head down. This approach was revived in certain schools in the United States three or four decades ago but is now largely being abandoned, whereas medical educationalists in this country have recently begun to implement it. It is easily seen, I think, that to fit some instruction in the history of medicine into such a 'regional' programme must involve teaching a very technological kind of medical history; the approach must be narrow and rigid, permitting the examination of detail rather than an appreciation of concepts, movements or trends. This approach may be contrasted with the more conventional approach to the history of medicine, or at least of scientific medicine, which is more in accord with the evolution of medicine itself. Scientific medicine commenced with an understanding first of structure, then of function, followed by structural and functional changes in disease, the integration of these with clinical experience and techniques, and finally an increasing understanding of aetiology and prevention. The conventional medical course followed history in considering all these subjects sequentially and more or less as separate disciplines, at least for purposes of teaching. This kind of approach thus has a logical form more appropriate to the teaching of history.

While the changes in the structure and content of the medical course to which I have referred probably make the inclusion of appropriate historical material more difficult, there is also today a lessened awareness of the significance of history amongst teachers, particularly the younger ones. It seems to me that senior teachers of medicine, whether clinical or academic, and however specialised, inevitably develop some historical perspective, whether they know it or not, with the passage of the years. Their teaching gains from this, even though the word 'history' may not be mentioned. Nonetheless it is true, as Crookshank rather ruthlessly put it, that 'a tolerant, and usually incorrectly verified allusion in the opening paragraph of an occasional address, that in itself gives renewed presentation to fallacies old before the pyramids were built, is — too often — the sole tribute paid by Modern Physicians to the History of Medicine'.[21] In part this stems from changes in the social and educational background of the students.[22]

Offset against the increasing difficulty in usefully fitting the history of medicine into a crowded curriculum is the attractive possibility of teaching a great deal of medico-social history in courses in community medicine. Epidemiology and demography are the basic sciences of community and social medicine (I should like to elevate social psychology to this category but as yet I dare not) and these are amongst the disciplines already mentioned as having a dimension of time. When community medicine becomes, as I fear it sometimes does, a 'soft' subject with more sociology than science, leavened with a zealous philosophy akin to the evangelism of nineteenth century religion, then an historical component would serve to return a few academic feet to earth and restore to the student a sense of reality. The inclusion of medico-social history in courses of community medicine also enables the ritualistic obeisance which must nowadays be made to the educational shibboleth of relevance. In Europe there is a tendency to link the teaching of medical ethics with teaching in the history of medicine; for some inexpli-

cable reason ethics seems to have disappeared from our medical curricula in
Australia, despite its relevance!

CONCLUSION

In this paper I have tried to indicate that the history of medicine and health
has an importance to the understanding of history, and especially social
history, beyond its significance as the story of technological development
and scientific achievement. The specialist historian in this field can therefore
contribute to the study of history and to its understanding in a variety of
ways. There are also ways in which specialties with a dimension of time, such
as epidemiology, may contribute to the investigation of historical material,
and an awareness of the epidemiologist's methodology will often have useful
applications in the historiography of communities of diverse kinds and in
differing enviroments.

Prince Henry Hospital, University of New South Wales

NOTES

[1] In recent years the best studies in this field have been by Lester King, for example
(1958) *The Medical World of the Eighteenth Century* (Chicago: Univ. Chicago Press);
(1963) *The Growth of Medical Thought* (Chicago: Univ. Chicago Press); (1970) *The Road
to Medical Enlightenment 1650–1695* (London: Macdonald). I am concerned here with
scientific rather than social philosophy.
[2] Cumston, C. G. (1926) *An Introduction to the History of Medicine from the Time
of the Pharaohs to the End of the XVIIIth Century:* with an essay on the relation of
history and philosophy to medicine by F. G. Crookshank (London: Kegan Paul, Trench,
Trubner and Co.), p. xxix.
[3] Gandevia, B. (1953) 'Shakespeare and Chaucer: Their Use of Medical Allusion in the
Story of Troilus and Criseyde', *Roy. Melb. Hosp. Clin. Rep.* 23, 9.
[4] Gandevia, B. (1977) 'A Comparison of the Heights of Boys Transported to Australia
from England, Scotland and Ireland c. 1840, with Later British and Australian Developments', *Aust. Paediat. J.* 13, 91. See also Note 17.
[5] Gandevia, B. (1976) 'Some Physical Characteristics, Including Pock Marks, Tattoos
and Disabilities, of Convict Boys Transported to Australia from Britain c. 1840', *Aust.
Paediat. J.* 12, 6.
[6] Elton, G. R. (1967) *The Practice of History* (Sydney: Sydney Univ. Press), pp. 5, 28.
[7] Gandevia, B. (1975) 'Socio-Medical Factors in the Evolution of the First Settlement
at Sydney Cove, 1788–1803', *J. Roy. Aust. Hist. Soc.* 61, 1.
[8] The word 'welfare' is used in *Australia 1938–1988,* Bull. No. 2 (1979) by Dan
Coward, Brian Dickey and H. G. Butlin to mean three different things, if I understand

them correctly (pp. 12, 13 and 18, respectively). 'Welfare state', a notoriously ill-defined term often with deliberately cultivated emotive overtones (*pro* or *con*), would give a total of four meanings, if it is conceded that the phrase is at all meaningful. Norma Townsend ('The Molesworth Enquiry: Does the Report Fit the Evidence?', *J. Austral. Studies* 1, 1977, 33) noted that 'reform' meant different things to most witnesses to the Molesworth Committee and to the priest, W. B. Ullathorne. Social historians, in particular, need to be more aware of these semantic problems than often seems to be the case.

[9] At the time this was written I did not have in mind (although I must have read) Elton's comment: 'I once heard [a sociologist] say that the study of the past is superfluous because a true understanding of the present, arrived at by sociological analysis, enables one to extrapolate and explain the past' (*op.cit.*, Note 6, p. 10).

[10] Briefly discussed in Gandevia, B. (1978) *Tears Often Shed: Child Health and Welfare in Australia from 1788* (Sydney: Pergamon Press), p. 120, but different views have been expressed by Brian Dickey and Michael Horsburgh in several papers published during the last decade.

[11] Gandevia, B. (1967) 'Medical History in Its Australian Environment', *Med. J. Australia* 2, 941.

[12] Gandevia, B. (1974) 'Mortality at Sydney Cove, 1788–1792', *Aust. N. Z. J. Med.* 4, 111.

[13] Gandevia, B. and Forster, F. M. (1974) 'Fecundity in Early New South Wales: An Evaluation of Australian and Californian Experience', *Bull. New York Acad. Med.* 50, 1081.

[14] Lancaster, H. O. (1952) 'The Mortality from Violence in Australia', *Med. J. Australia* 2, 649.

[15] Gandevia, B. (1978) 'Malingering in the Penal Era: Its Epidemiology and Social Implications, with Contemporary Observations by Dr. James Stuart', in *Festschrift for Kenneth Fitzpatrick Russell*, ed. H. Attwood and G. Kenny (Melbourne: Queensberry Hill Press), p. 59.

[16] Gandevia, B. (1976) 'The Epidemiology of Dysentery at Norfolk Island 1840–1843', Proc. Ann. Mtg. ANZSERCH, Brisbane, p. 198.

[17] Gandevia, B. (1975) 'The Height and Physical Characteristics of Convicts Transported to Australia *c*. 1820–1850', Proc. Ann. Conf. Aust. Assoc. Hist. and Phil. Sci., Sydney; see also Note 5, *supra*.

[18] The most comprehensive source for information on the infectious diseases in Australia is D. Gordon (1976) *Health, Sickness and Society* (Brisbane: Univ. Qld. Press). Some further references of historical interest are in Gandevia, B. (1957) *An Annotated Bibliography of the History of Medicine in Australia* (Sydney: Australasian Med. Publ. Co.), and in Lancaster, H. O. (1964 and 1973) 'Bibliography of Vital Statistics in Australia and New Zealand', *Aust. J. Statist.* 6, 1964, 33 and 15, 1973, 1.

[19] I am grateful to the Editors, *Australia 1938–1988 Bulletin*, for permission to use material published in Bulletin No. 3, 1979. M. M. Postan (1968) 'Fact and Relevance in Historical Study', *Hist. Stud.* (Melb.), 13, 411, looks at analogous problems from the historian's viewpoint.

[20] I claim no special expertise or experience in teaching the history of medicine. I have given more seminars in this subject outside medical faculties than in them. The role of medical history in medical education and the means of teaching it have been the subject

of many notable publications in recent years: see, for example, a series entitled 'View-points in the Teaching of the History of Medicine' in issues of *Clio Medica* for 1975; Blake, J. B. (ed.) (1968) *Education in the History of Medicine* (New York: Hafner); Poynter, F. N. L. (ed.) (1969) *Medicine and Culture* (London: Wellcome Inst. Hist. Med.). Other recent books offer suggestions for courses and syllabi.

[21] *Op. cit.,* Note 2: p. xviii.

[22] Garlick, H. W. (1972) 'The Winds of Change', *Med. J. Australia Supplt.* 1, 17. A student referring correctly to a collection of dilated and tortuous veins as a 'caput Medusa(e)' added that Medusa was the Italian who first described the lesion. Is it possible to cope with this level of ignorance by teaching history and ethics? Some 'educationists' (a curious word) would wonder if it matters.

JARLATH RONAYNE

SCIENCE POLICY STUDIES: RETROSPECT AND PROSPECT

Study of the relationship between science, technology and society (SSTS) is an element of the discipline of History and Philosophy of Science that has experienced remarkable growth in recent years. Undergraduate and graduate courses have proliferated, research units have been set up, specialised societies and journals established and an enormous body of literature created. There is no doubt that the stimulus for these developments in SSTS was provided by the massive governmental involvement in scientific and technological activities that took place in the 1950s and 1960s. By the mid-1960s, the need for disciplined inquiry into the social, economic and political implications of the growth of science and technology in advanced industrial societies was obvious, and SSTS as a new subdiscipline developed because the established disciplines were unequal to the task.

In an attempt to bring some order to the diverse areas subsumed under the heading of SSTS, Spiegel-Rösing has divided the sub-discipline into two fairly distinct branches — *social studies of science* and *science policy studies*. Whereas she sees SSTS research as a whole deriving its justification for support in terms of its potential for the solution of practical problems, there is little doubt that science policy studies can be justified in these terms to a far greater extent than social studies of science. As Spiegel-Rösing says,

World War Two is a crucial turning point in the changed relationship of science to power. It is this very change that led to the largely dominant political attitude of *laissez-faire* toward the development of science and to massive government support of science: in turn this trend led to the need for control and direction of science with the concomitant need for instrumental knowledge of control.[1]

Science policy research derives its justification from the need to provide the instrumental knowledge of control. It is this branch of SSTS that will be my main concern.

The appropriate form of training in this branch of SSTS is simply an element of the larger question of the appropriate form of training in the sub-discipline as a whole; the problems that are raised are similar. Is science policy studies an area that needs a new kind of professional, or should the necessary expertise be provided as an adjunct to professional studies in the

R. W. Home (ed.), Science Under Scrutiny, 99–121.
© *1983 by D. Reidel Publishing Company.*

natural or social sciences? Can the necessary skills and concepts of policy analysis and management be taught in a year or so? Are people trained in this manner capable of providing for the decision-makers accurate informed advice on scientific and technological policy matters? If training in science policy studies is to be imparted merely as an adjunct to other professional studies, will this exacerbate the problems of identity and intellectual rigour that are said to exist in the whole area of SSTS?[2,3,4] These questions are not yet resolved.

Whatever the approach adopted towards the training of science policy analysts, however, the desirability of such training is generally not open to doubt. Nor is there any real doubt as to the need for research to be carried out on science policy questions. The social function of research (and training) in science policy is to:

avoid the amateurism, needless conflict, inefficiency and simplistic approaches to contemporary problems resulting at least in part from analysis and policy decisions made and implemented by people who do not understand the interaction between science and technology or its relationship to social change.[5]

We are now, it seems at a stage of development in science policy research where it is appropriate to ask what has been achieved, and here there are signs of pause, of ongoing evaluation. Spiegel-Rösing provides us with an elegant review of the state of the art. From her analysis it is evident that there are certain deficiencies in the area of science policy research. It is highly fragmented; there is a lack of consensus between separate methodological and conceptual approaches; there is a lack of comparative research; and there is a preponderance of 'individual pieces of issue-oriented research'.[6] Haberer goes further than this. In his review of the progress of science and technology policy over the past twenty-five years, he suggests that

there is little evidence that an intellectual effort is underway to create a more rigorous comparative theory or even a unifying framework. To a considerable degree we have a body of literature in search of a field.[7]

If science policy research workers can find little comfort in all this, the customers — the science policy makers, the research managers and so on — can find even less. The central questions of science policy concern the organisational frameworks that are appropriate for the management of research and development within specific institutional settings, and the allocation of scarce resources in the aggregate or between fields of scientific endeavour. According to Spiegel-Rösing, science policy researchers have provided no

answers to these questions. Their customers are advised that sweeping generalisations about science policies and strategies of research support cannot
and should not be made, and that resource allocation must remain a process
of 'muddling through'. Linear models of innovation which would ease the
problem of resource allocation are swept aside, and perils in external direction and control of scientific activity are continually highlighted. The obvious
dangers to the development of the field are apparent to Spiegel-Rösing. 'It
is not easy to question the user interest', she suggests, 'and at the same time
avoid undermining financial support for one's own research.[8]

This is a very pessimistic view, and I looked for something more positive
in Schmandt's more recent review of science policy research, in which he
tried to identify from the literature some organisational principles that
might be graced with the title of *science policy constants*.[9] In order that
old battles would not have to be fought over and over again, these science
policy constants would describe a few central and continuing issues that
would be kept in mind when decisions on funding, organisation and linkage
to other policy goals were being made. Schmandt gave two concrete examples.

The first of Schmandt's science policy constants referred to the need for
institutional differentiation in the performance of research and development
(R&D) tasks, and the consequent need to examine policies, regulations and
funding practices for their impact on the different institutions performing
the research. This may appear to be self-evident, but it has not always been
perceived by planners. The disastrous results arising from the imposition of
R&D tasks on or their encouragement in institutions that are organisa-
tionally or structurally incapable of performing them can be demonstrated by
examples such as NASA's Sustaining Universities Program and the National
Science Foundation's RANN Program.[10]

The Sustaining Universities Program (SUP) was a NASA attempt to relate
university research to the needs of society, seeking to link portions of the
scientific community into a research system whose output would be information and technology of immediate relevance to problems of national
concern. NASA had at the time of SUP's inception, and continues to have, a
close relationship with the universities, to whom it contracts a great deal of
work on aeronautics and space science. In 1962, the Agency decided to encourage the universities to undertake research applied to broader, more
'national' needs rather than those specific to the space programme. Funds
were set aside for NASA fellowships, research grants and buildings, and the
terms and conditions under which the universities would receive support
under the scheme were laid out in *memoranda of understanding*, signed by

the participating university administrators. One of the more important clauses in the memoranda was the requirement that the universities should seek ways of diffusing the results of space-related research by putting the funds to work in advancing development and solving technical, economic and social problems in the universities' own regions and in the nation as a whole.

By certain standards the SUP was a success; about $200 millions were allocated to the participating universities between 1962 and 1970. However, the multidisciplinary aspect of the SUP was not taken seriously by the participants; they perceived the grants as institutional support in the conventional sense, that did not require innovations in the administration of research. The memoranda were ignored and the programme goals set out within them were not achieved, because the universities failed to respond to the demand for multidisciplinarity and technology transfer. The programme was terminated in 1970 and in the post-mortem the blame for its failure was laid squarely at the doors of the universities. If blame could be attached to NASA it was simply in terms of its naïveté in expecting the universities to change. As Blankenship has put it:

The NASA director undoubtedly believed that anything was possible. He was able to lead America to the Moon, but he could not change the university.[11]

The leverage exercised by NASA — money, buildings and agreements — was insufficient to effect changes in the universities' operations. This was an attempt to change the universities' traditional habits from the outside, to influence their priorities. But it was left to the people inside the system to choose the manner in which the 'reform' was to proceed. The SUP's clients were in an environment that worked against the aims of the Program.

The National Science Foundation's Research Applied to National Needs (NSF-RANN) programme was established in 1971. It had its origins in the Interdisciplinary Research Relevant to Problems of Our Society (IRRPOS) programme, established two years previously as a response to the amendments to the NSF Act of 1968 which authorised it to fund applied research, and secondly, as a response to the political climate of the time. Like the SUP, RANN's objective was to induce the universities to undertake research of relevance to the perceived needs of American society. It failed, and was discontinued in 1976.[12] RANN is an important episode in the history of American science policy. It was an attempt by an agency that traditionally had no mission other than the support of basic research to extend its operations into applied areas of research using its traditional methods of proposal pressure and peer review to identify priority areas of research. Schmandt's first science policy constant was unrecognised.

The lessons of SUP and RANN should not be lost on the decision-makers, especially in the light of present preoccupations with so-called national programmes aimed at solving difficult, multidisciplinary problems of our time, in relation to energy, food, population and resources, for example.[13] So far, no obvious institutional structure has emerged where research of this nature, requiring many talents and resources, is best performed.

Schmandt's second science policy constant — 'the need to recognise that multiple funding sources are not wasteful as long as each agency has meaningful review systems for judging the quality of all proposals'[14] — is less persuasive than the first. It rests on the belief that the availability of many sources of funding for R&D, not necessarily co-ordinated, is a protection against judgemental errors in reaching decisions about allocations; uncertainties about the outcome of proposed research programmes cannot be eliminated from such decisions and hence a strategy of attacking the same problems from different points of view is held to be desirable. This is a classical pluralist approach to science policy, and in advocating it Schmandt is expressing a personal preference. There are, as I shall shortly indicate, other organisational forms that might be more appropriate in different economic, social and political contexts. The definition of a pluralist organisational mode as a science policy constant is, to my mind, highly problematic. To be sure, in certain circumstances this organisational mode can be very attractive and appropriate. In a very wealthy nation, where the development of criteria of scientific choice and priority setting in research funding are not important elements of public policy, unco-ordinated pluralistic funding mechanisms may be socially, economically and politically acceptable. Under conditions of national emergency, multiple approaches to the same problems may be the only practical way to proceed. The decision to press ahead with the simultaneous development of the three possible methods of isotope separation in the Manhattan Project was taken because the United States could afford the expense involved and time was of the essence. In peacetime, and under conditions of greater financial restraint and accountability, such an approach cannot be regarded as a universal science policy constant. This applies *a fortiori* to basic research, where research problems are rarely of such importance that their investigation requires multiple approaches.

Thus, of the two science policy constants isolated by Schmandt from the science policy literature, the first, that different R&D tasks require different institutional settings and that this should be taken into account in any science policy initiatives, is a valid one. The second, that pluralistic forms of research support are the most appropriate for R&D is, to my mind, less compelling. I suggest that there is no universal constant in the fundamental area of science policy that refers to the appropriate organisational structure which governments might adopt in implementing their policies for R&D. At least four organisational models of science policy machinery are in operation in different countries of the world. If we accept that science policy machinery is both necessary and desirable, then that itself can be a constant; but it contains within it the possibility of a number of organisational forms that could be classed as variables. Each of these variables can be illustrated by one or more concrete examples. By techniques yet to be developed — science indicators, perhaps — it may in the future be possible to evaluate the efficacy or otherwise of each organisational model. In this way we may one day be able to overcome the criticisms that have been voiced about the lack of comparative research in science policy, and answer the charge that results that are valid for more than one institution, one research area, or one sample of research scientists are rarely achieved.[15] In the following section I describe the four organisational models for governmental R&D, and some comparative research in science policy, which will, I hope, show that the literature on science policy is not as sterile as Haberer suggests and that there have, in fact, been concrete achievements relevant to the problems of the policy makers.

SCIENCE POLICY MACHINERY

Certain key elements can be identified in the machinery that many countries have adopted for the management of their R&D support and operational systems. Governments perform R&D in their own departments and agencies, they support it in universities, research institutes and colleges, and they promote it in the private sector. Co-ordination and control is exercised to a greater or lesser degree by independent advisory bodies, interdepartmental committees, interministerial committees and special ministers of state. Depending on the organisational model that has been adopted, some or all of these elements will be present in any national science policy system. It is important to note, however, that it is very difficult to point to examples that match the ideal models exactly, and there is still some difficulty over precise definitions.

The Centralist Model

There has been, in the past, some confusion in the literature as to the precise definition of a centralised science policy system. The Canadian Senate Special Committee on Science Policy (the Lamontagne Comittee)[16] defined a centralist system as one in which the bulk of long-range governmental research and development is carried out within a single agency or department of state, the other departments confining themselves to short-range research geared specifically to their missions. The British system in force between 1947 and 1964 was defined by the Lamontagne Committee as centralist (see Figure 1). The Department of Scientific and Industrial Research (DSIR), an independent research council, was responsible for all long-range research which it carried out on its own initiative or (theoretically at least) at the request of the administrative departments. The Advisory Council for Scientific Policy, reporting, like DSIR, to a Minister-without-Portfolio (the Lord President of the [Privy] Council) advised the latter in the exercise of his responsibilities for the formulation and execution of government scientific policy.

Brooks, on the other hand, identified a centralised science policy system as one in which the whole of the governmental research effort, long- and short-range, is carried out within a single department of state.[17] Nowadays it is the notion of central planning of scientific research and centralised distribution of financial resources to the government agencies and departments that are responsible for R&D that is generally associated with a centralised model. In France, for example, the Minister with responsibility for civil R&D negotiates funds for the programmes of the ministries which carry it out. R&D expenditures are therefore isolated in the annual budgetary process and although the appropriations from the Ministry of Finance are made directly to the individual ministries, their R&D budgets are fixed as a result of the pre-budget negotiations with the Minister. Research programmes are carried out within the framework of a five-year national plan, the scientific research elements of which are formulated within the bureaucracy with the advice of an advisory council. Thus, resource allocation is centralised since the overall science budget is fixed by a process that sets the science budget of one ministry against that of all the others and vice versa, and research is carried out within the context of an overall national plan. Such centralised planning and resource allocation mechanisms can be combined with concentration of research activity within the government sector. In France, the Centre National de la Recherche Scientifique (CNRS),

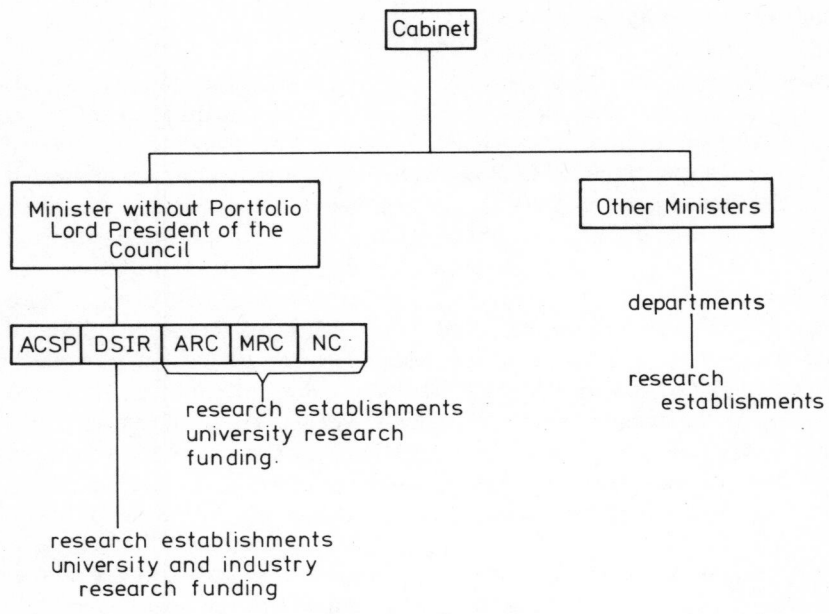

Abbreviations:

Advisory Council for Scientific Policy (ACSP)
Department of Scientific and Industrial Research (DSIR)
Agricultural Research Council (ARC)
Medical Research Council (MRC)
Nature Conservancy (NC)

Fig. 1.   Science policy machinery in the United Kingdom, 1947—1964.

with a staff of 22000 and research budget in excess of $600 millions, performs the bulk of French governmental R&D.

The Concerted-Action Model

In the concerted-action model, co-ordination of governmental R&D is achieved by a minister, usually a Minister of Science Policy, who acts with the concurrence of all ministers who have responsibility for the performance of R&D. The co-ordinating minister, who is usually provided with a secretariat, prepares the decisions to be finalised by interministerial committee, a procedure which, it is believed, ensures that the interests of the different policy sectors are safeguarded and that science policy looks beyond the boundaries separating the various ministries. The Minister of Science Policy is usually given the power to assess the R&D programmes drawn up by the other ministries and is given the policy instrument of a science budget. The Dutch, Belgian and Canadian science policy systems illustrate the concerted-action model of science policy.[18] (See Figure 2.)

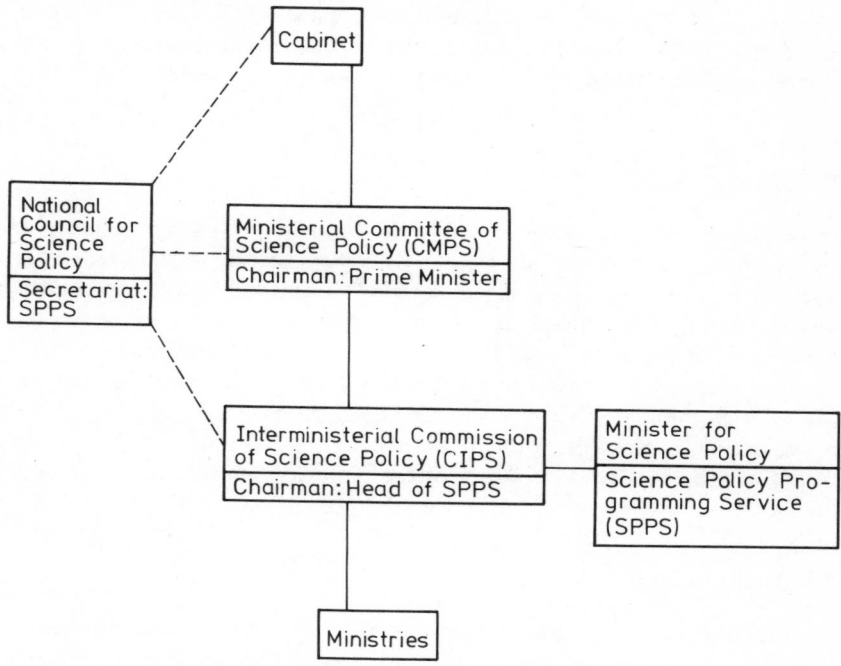

Fig. 2.   Belgian science policy machinery (concerted-action).

The Co-Ordination Model

In the co-ordination model the various ministries and agencies act independently but there is a general awareness that their research activities should be co-ordinated. Post-Rothschild Britain provides us with an example of a country that has apparently adopted a co-ordination model. In the changes that were put into effect as a result of the Rothschild Report[19] the government

ministries were given responsibility for 'acquiring' the R&D they needed for
the fulfilment of their missions. Decisions were related to departmental objectives and did not fit into any nationally determined policy for science and
technology. Adjustments have been made to this system, however. In 1976 a
co-ordinating committee of departmental Chief Scientists and Permanent
Heads was set up. Further co-ordination is supposed to be achieved within
the Cabinet Office where the Chief Scientist of the Central Policy Review
Staff has widespread powers to inquire and advise. Finally, the Advisory
Committee on Applied Research and Development (ACARD), also established in 1976, is designed to improve the interface between government and
non-government organisations in matters related to applied R&D. (See
Figure 3.)

The Pluralist Model

In the pluralist model, best illustrated by the United States, financial resources
are assigned to each sector as a whole (defence, health, agriculture, energy
and so on). The needs of the sector for R&D are determined in competition
with other capital investments, operational needs and servicing needs, and the
R&D is either contracted out to other agencies or performed intramurally.
Usually, each sector operates independently of the others and priority-
setting is an internal affair. Even in the highly pluralist U.S. system, limited
co-ordination is possible through the Office of Science and Technology
Policy within the Executive Office of the President, whose Director is also
the President's chief scientific advisor. (See Figure 4.)

Comparative Research in Science Policy

The four science policy organisational models are present to a greater or lesser
extent in the executive and advisory systems of almost all of the OECD
member countries and other advanced industrial nations. Each has strengths
and weaknesses, and the models themselves or elements thereof can be subjected to comparative analysis if we can agree on the basic definitions. A
wealth of standardised data is available from UNESCO, OECD and other
sources upon which comparative studies can be based. In countries like
Australia, where science policy machinery is still in a state of evolution, such
studies could provide invaluable insights into the appropriate final machinery
that should be adopted, though of course the social, economic and cultural
differences that exist between nations must always be borne in mind.

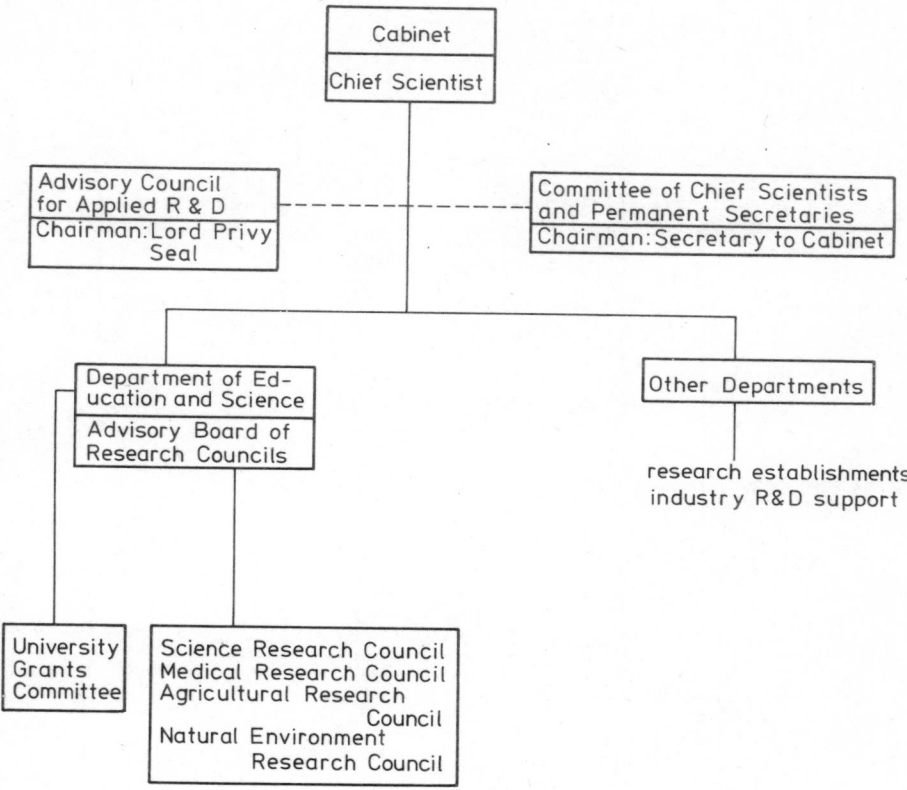

Fig. 3. Science policy machinery of the United Kingdom (co-ordination), 1981.

A recent study by Brickman and Rip[20] demonstrates the value of such comparative research and indicates that one of the main problems of science policy research, the lack of comparative data, is at last being tackled. In a study of the influence of scientists in decision-making processes in government, Brickman and Rip examined in detail the principal scientific advisory bodies of three countries — the United States, the Netherlands and France. Their conclusions should be of interest to the members of science policy advisory bodies everywhere, and to the governments they serve. In the United States, the President's Science Advisory Committee (PSAC) was established in 1957 and eventually became the independent advisory element of a science policy co-ordinating system that included the Office of Science

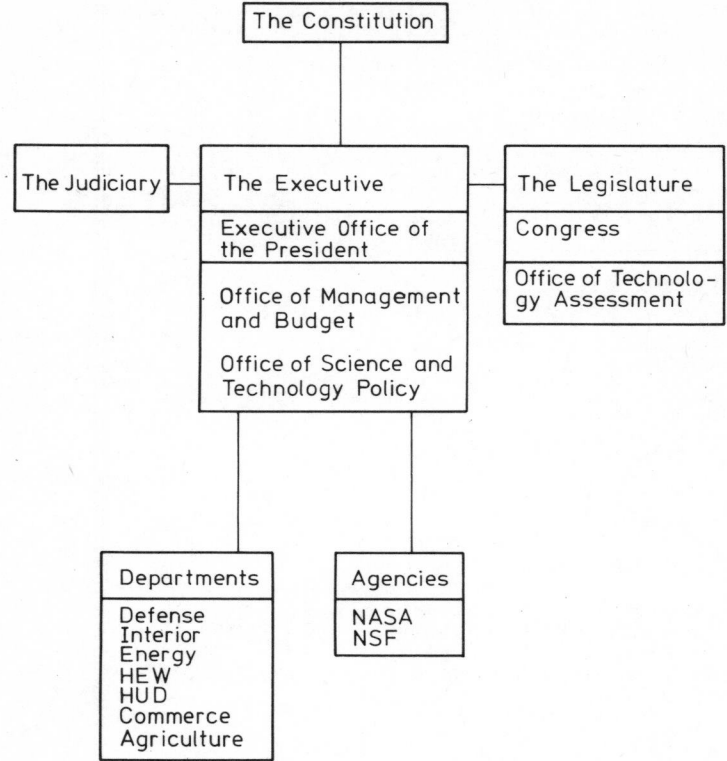

Fig. 4.   United States of America: federal research and development advisory and
executive system (pluralist).

and Technology and the Federal Council for Science and Technology. In
France, the Consultative Committee for Scientific and Technological Research
(CCRST) was established in 1958; in the Netherlands, the Advisory Council
for Science Policy (RAWB) was created in 1966. In each case, the advisory
councils, after an initial period in which they played an important and influential role in the scientific affairs of their governments, went into a long
period of decline. The PSAC was abolished by President Nixon in 1973 and,
while RAWB and CCRST still exist, their influence on decision-making
processes is now negligible. A cursory look at the advisory councils of other
countries shows a similar pattern, though the study by Brickman and Rip

remains the only extended comparative analysis that has been carried out. For example, the British Advisory Council for Scientific Policy (ASCP) has been abolished, and the Science Council of Canada, the German Science Council (Wissenschaftsrat), the Belgian National Council of Science Policy (CNPS), the Swedish Government Research Advisory Board and the Swiss Science Council have all gone into a decline.[21]

The United States Committee and the French and Dutch Science Councils were victims of their own success. Each had lobbied for a coherent, reasoned, institutionalised approach towards governmental intervention in science and technology. The Councils were set up during a period when the research and development expenditures of the countries concerned were growing, and before regular institutional mechanisms were perfected to handle the complex and esoteric questions raised by such heavy involvement by governments in R&D. The institutionalisation of science and technology in the bureaucratic processes of government removed the Councils from the corridors of power and their decline set in. In the United States, the PSAC fell victim to the expansion of central administration within the Department of Defense and the R&D-oriented agencies such as NASA and the Atomic Energy Commission, and to the proliferation of advisory and policy-making staff within the White House itself. In France, the influence of CCRST was undermined by the creation within the bureaucracy of the Ministry of Research and Industry to co-ordinate governmental R&D. In addition, an interministerial committee was formed (Comité Interministériel de la Recherche Scientifique et Technique), the task of which was to prepare for final government action the country's major options in science policy. In the Netherlands, the creation of a Ministry of Science Policy with the support of a Directorate-General for Science Policy spelled the end of RAWB as an influential body. Extending the analysis, the British Ministry of Technology, the Canadian Ministry of Science and Technology, the Belgian Ministry for Science Policy with its powerful Science Policy Programming Service (SPPS), and the West German Ministry of Research and Technology all rendered the respective Science Councils ineffective.[22]

Now the institutionalisation of science policy within the machinery of government was not the only factor identified by Brickman and Rip as contributing to the decline of the advisory councils, but it was the most important. The comparative study provides valuable background information for countries whose science policy machinery is being established or is in transition. My only complaint is that, in defining the types of science policy machinery that they found in the three countries they studied, they followed what Price has called the Messianic Tradition[23] and devised organisation models of their own (atomistic, low co-ordination and high co-ordination) that neither refined nor extended the existing definitions which they cited but, inexplicably, decided not to use. Not until science policy researchers can agree on fundamental definitions will science policy research be able to provide that 'instrumental knowledge of control' which is needed no less pressingly now than it was in the 1960s, when this branch of learning was given its first real impetus.

THE AUSTRALIAN PERSPECTIVE

In this final part of my paper I shall discuss the situation in Australia, bearing in mind that my previous discussion of the function of science policy research refers to organisational matters in the civil sector only. I begin with a quotation from Alvin Weinberg's recent review of the *Science and Government Report International Almanac – 1977,* which opens thus:

Malcolm Fraser, when Minister of the Australian Department of Education and Science, is quoted in this book as having denied that a science policy was definable let alone worth talking about. Has the lengthy debate in priorities in science that graced *Minerva's* pages in the 1960s and still reverberates in the academic science policy community been much ado about nothing, or does the new Prime Minister of Australia display ignorance about the real questions that have been defined, have been talked about – and, perhaps, in the talking have even influenced what has happened?[24]

I do not know whether this is a reflection on Mr. Fraser, on Weinberg and his criteria for scientific choice, or on the author of the 'Australia' section of the *Science and Government Report International Almanac,* but it certainly shows that for commentators on Australian science policy, memories are long. Mr. Fraser made his statement in 1969. It is usually John Gorton, former Minister for Education and Science, and former Prime Minister, who receives the 'wooden spoon' from science policy analysts for his immortal words, uttered in 1968 and quoted by science policy scholars to this very day:

I don't know what a science policy is. The critics want an overall advisory committee to allocate funds but I don't see the need for an advisory body. The committees are only a group of individuals pushing the barrow for their own disciplines.[25]

But the central question posed by Weinberg is a valid one. Have our decision-makers derived any benefit from the science policy literature? And, in line with my main preoccupation in this paper, have we in Australia learned from overseas experience in science policy matters?

There are elements in the Australian government's science policy machinery that are similar to those of other countries, but the differences between our operational and advisory systems and those in other advanced industrialised nations are really more striking than the similarities.[26] Broadly speaking, the Australian governmental research and development system is pluralist. (See Figure 5.)

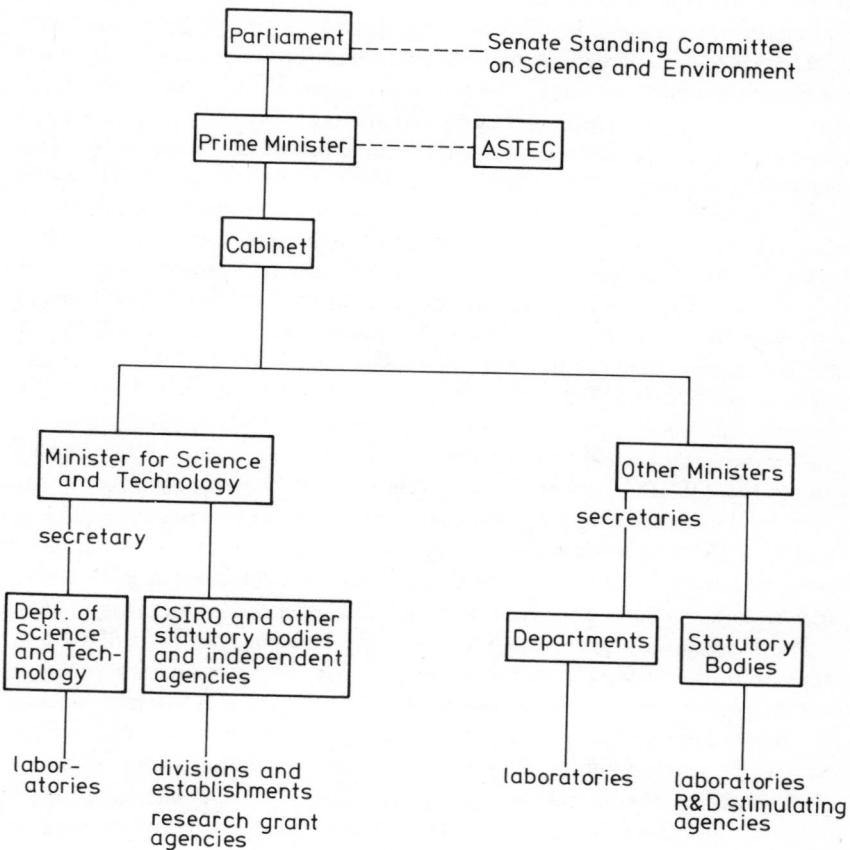

Fig. 5. Australian science policy machinery, 1981.

Each department of state is responsible for procuring the research and development it needs for the efficient fulfilment of its mission, and some departments maintain laboratories whose function is to service departmental research needs. Departmental research and development personnel are employed under the terms and conditions of the Public Service Board; laboratory directors report directly to the permanent head of the Department who, in turn, reports to the Minister. There is no 'science budget', and research expenditure proposals are not usually separated out from the main departmental expenditures in the budgeting process.

Certain agencies and laboratories have been created for the performance of research of a more general nature. These statutory corporations, of which the Commonwealth Scientific and Industrial Research Organization (CSIRO) is the most famous, are established by law and report directly to a designated Minister, not to the permanent head of an administrative department. This organisational form will be familiar to students of science policy. It derives from British experience and is in accordance with the principles laid down by Lord Haldane in 1918, when the British research council system was established and the Department of Scientific and Industrial Research became the model for similar developments in Canada, South Africa, India, New Zealand and Australia. CSIRO, the DSIR's Australian counterpart, is the largest research organization in the country. Taken together, the CSIRO and the Defence Science and Technology Organisation carry out more than fifty percent of all governmentally funded research and development. The proportion of research carried our by government agencies is one of the highest among the OECD member states.[27] Thus, while the organisation of the governmental research and development system can be described as pluralist, there is a heavy concentration of research performance in the government sector and, within that sector, research is further concentrated in two very large research organisations, one of which has independent statutory status.

The Minister for Science and Technology is responsible for CSIRO and other statutory research agencies, but he has no co-ordinating function over the governmental research system as a whole. The Department of Science and Technology is, to my mind, a unique Australian invention. It has both operational responsibilities and, according to the administrative orders, a role in assisting the government in the formulation and implementation of its science and technology policies. The advisory function has never been an important part of its portfolio, however.

The main source of advice to the government on scientific and technological matters is the Australian Science and Technology Council (ASTEC), a statutory advisory body reporting directly to the Prime Minister. ASTEC has no executive power, but is a source of both confidential and public advice to the Prime Minister. It can be, and is, invited to comment on major scientific proposals emanating from departments and statutory bodies.

These are the basic elements of Australia's science policy machinery. The co-ordination of research activities that an external observer might expect to find, given the existence of a top-level advisory body and a Department of Science and Technology with terms of reference that include the provision of advice on policy formulation and implementation, is almost non-existent.

We may ask why it is that Australia's science policy machinery has developed in this way. Numerous papers have been published on developments that have taken place since 1966, and it is beyond the scope of this paper to review the entire period,[28] so I shall concentrate on developments since 1972. The Labor Party platform in that year foreshadowed the appointment of a Minister for Science who would have no departmental responsibilities, but who, together with an advisory council, would formulate policy recommendations for submission to Cabinet.[29]

The Prime Minister's interpretation of the platform was, however, less clearcut than this. Mr. Whitlam suggested, two years before he took office, that the Minister for Science, together with an advisory science council, 'would formulate policy recommendations for submission to Cabinet and would seek *through the operational arm of his portfolio — CSIRO and comparable organisations* — and through the advisory council, the implementation of policy'.[30] Whereas the platform effectively foreshadowed the creation of a minister for science policy, the future Prime Minister's statement did not. The public service contingency plan which was put into effect when the Labor Party won the election in 1972 followed the guideline laid down by Mr. Whitlam, and the Minister for Science, Mr. W. L. Morrison, acquired an administrative department.

One of Mr. Morrison's first executive decisions was to disband the top level Advisory Council for Scientific Policy, set up by the previous government earlier in 1972. In attempting to formulate the terms of reference, membership categories and methods of appointment of the Labor Government's successor to this body, Mr. Morrison initially engaged personal 'science policy' advisors drawn from the ranks of the Labor Party. He then sought the advice of ministerial and departmental colleagues, of the scientific community and, finally, of the Organization for Economic Cooperation and Development (OECD). The OECD examiners proposed a concerted action model of science policy machinery for Australia with the following key elements:[31]

(a) a Minister for Science without operational responsibilities;

(b) a Ministerial Committee chaired by the Minister for Science which would have a co-ordinating function and exercise its powers through control over a 'science budget';

(c) a powerful advisory council, possessing a small secretariat, which would assist the Ministerial Committee and report to the Prime Minister through the Minister for Science. Thus, 'on all matters of detail and substance the Ministerial Committee would be assisted by an Advisory Council for Scientific and Technological Policy, which would prepare the work of the Ministers and work out the details of the national science policy in consultation with appropriate experts.

The government's decisions on its science policy machinery, published in January 1975,[32] bore a superficial resemblance to the OECD recommendations. The Ministry of Science would continue, a Ministerial Committee would be set up and an advisory council (ASTEC) established. However, the Minister for Science would retain his operational responsibilities, exercised through statutory corporations and the Department of Science; the latter would also act as 'agent' for ASTEC. ASTEC would report to the Ministerial Committee and this Committee would also be advised by interdepartmental committees on all matters referred to ASTEC and on matters considered inappropriate for reference to ASTEC. The structure of these interdepartmental committees was not revealed in the White Paper, but it is clear that by their retention the existing powers of the statutory corporations and departments in scientific and technological matters would be preserved and the powers of ASTEC curtailed. And since there was no mention in the White Paper of a science budget, the Ministerial Committee and ASTEC would be denied the prime co-ordinating instrument in any case.

In effect, therefore, the Labor Administration rejected the fundamental recommendations of the OECD examiners, as had Mr. Fraser, the then Shadow Minister for Science, who dismissed the OECD recommendations because 'examiners from overseas might not necessarily be able to make the best judgements about these matters'.[33]

With the change of government in November 1975, the terms of reference and membership categories of ASTEC were reviewed and in 1976 a new ASTEC came into operation, reporting to the Prime Minister and with a membership that, apart from the inclusion of a political scientist and a trade unionist, reflected the government's original thinking in 1972. Although the same departments continue to send representatives to ASTEC meetings, there is no Ministerial Committee, nor any other co-ordinating element.

But there is, as yet, no real stability in the arrangements. In 1977, ASTEC reported for a short time to the Minister for Science, before reverting to the original arrangement in 1978 after intense lobbying. And waiting on the sidelines are the recommendations for re-organising Australia's science policy machinery prepared by the Science Task Force of the Royal Commission on Australian Government Administration.[34]

On the grounds that the submissions being made to it from the government's scientific sector revealed a widespread sense of frustration, the Royal Commission on Australian Government Administration set up a task force to examine the conduct and co-ordination of scientific work being carried out, financed, or supported by the government, to review the present organisation of scientific research, and to suggest appropriate new arrangements. Of the eight members of the task force, five, including the co-ordinator, were current or former employees of CSIRO, two were senior academics, and one was an industrialist. I shall confine myself here to its recommendations concerning Australia's science policy machinery.

The task force based its science policy — and I use the word loosely, for it displayed a distinct antagonism towards the very notion — on the theories of Robert Merton, Michael Polanyi, and Sir Karl Popper. It was argued that if the national scientific effort were organised according to the principles espoused by these authors, by granting as much autonomy as possible to the governmental scientific agencies, then most of the problems perceived by the task force would be mitigated, if not eradicated. For, so the task force maintained, it was an excess of uniformity, centralisation and rigidity that lay at the root of all the problems. The organisational framework proposed by the task force had much in common with that which existed in the United Kingdom between 1947 and 1964 (see Figure 1) and is shown in Figure 6.

The science policy literature, apart from that which agreed with the theories of Polanyi and Merton, might never have existed. The task force abhorred in all its forms the notion of 'centralism', but failed to define what it meant by that term. An explicit science policy seemed to be the very embodiment of centralism and was condemned as a fashion which most of the world's developed countries had adopted — though some unspecified nations were said to have quietly dismantled the machine and gone back to decentralised systems more tolerant of diversity![35] There is no doubt that it was centralisation of resource allocation and planning that the task force feared, and its recommendations, which included the abolition of the Department of Science, reflected this fear. ASTEC would be retained, yet it is interesting to note that the establishment of the Advisory Council for Scientific Policy, a cornerstone of Britain's postwar science policy machinery

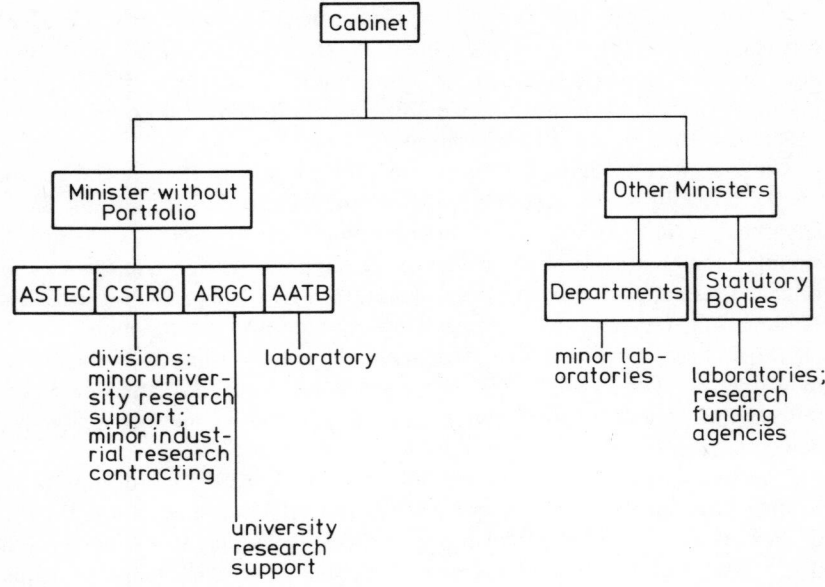

Abbreviations:

Australian Science and Technology Council (ASTEC)
Commonwealth Scientific and Industrial Research Organization (CSIRO)
Australian Research Grants Committee (ARGC)
Anglo-Australian Telescope Board (AATB)

Fig. 6. Science Task Force Proposals, 1975.

which had so much in common with that proposed by the task force, was considered by Gummett and Price[36] to be an attempt by the British Government to centrally plan civil research and development. Had it had a sense of history, the task force would, no doubt, have been comforted by the fate of the Advisory Council for Scientific Policy. Though it never perceived ASTEC as a central planning device, the possibility that it might become one would have been shown to be remote if the British experience was anything to go by.

Most of the science task force recommendations were accepted by the Royal Commission into Australian Government Administration but, because of the change of government that took place in November 1975, they were never put into effect. No attempt has since been made to re-organize the Australian governmental R&D system as a whole. Piecemeal adjustments have been made to elements of the system but no real effort has been made to co-ordinate these elements. There is no determination at governmental level to pursue a particular science policy model. Despite the fact that the pluralist model which is being pursued — albeit unwittingly — is clearly inappropriate in a country like Australia with limited appropriations to R&D, there has been no serious effort to examine alternative models. The governmental research effort is organized in such a way that, in general, research rather than development is emphasised. Until relatively recent times there was little evidence that government thinking on its intramural research effort, and its science and technology policy in general, was influenced in any way by the science policy literature. But this situation may change. Since its establishment in 1977 as a permanent advisory body, the influential Australian Science and Technology Council has submitted to the Cabinet a series of highly professional reports on important issues in science and technology policy. The Department of Science and Technology now produces an annual Science Statement which analyses the nation's gross expenditure on research and development in such a way as to provide invaluable background information on decisions on resource allocation and the maintenance of a balanced scientific effort. In ASTEC, in the Policy Division of the Department of Science and Technology, and in the growing number of universities with policy-oriented history and philosophy of science departments and STS programs, Australia now possesses a significant SPS infrastructure. Only time will tell whether or not the 'instrumental knowledge of control' that these agencies produce will be heeded by the politicians.

University of New South Wales

NOTES

[1] Spiegel-Rösing, I. (1977) 'The Study of Science, Technology and Society (SSTS): Recent Trends and Future Challenges', in Spiegel-Rösing, I. and de Solla Price, D. (eds.), *Science, Technology and Society: A Cross-Disciplinary Perspective*, London: Sage Publications, pp. 7–42.
[2] Miller, C. M. L., Moseley, R. and Ford, G. (1980) *The Impact of Science Technology and Society Courses in Higher Education*, Sussex: University of Sussex Education Area Occasional Paper No. 7.

3 Szyliowicz, J. S. (1977) 'Education for Science and Technology Policy Analysis: Problems and Prospects', in Haberer J., *Science and Technology Policy*, Lexington: D.C. Heath, pp. 143–149.

4 (1980) *Science Studies: A Report to the Nuffield Foundation*, London: The Nuffield Foundation.

5 Szyliowicz, *op. cit.*, p. 146.

6 Spiegel-Rösing, *op. cit.*, p. 35.

7 Haberer, J. (1977) 'Introduction: An Agenda for Science and Technology Policy: The Road Not Traveled – Yet', in Haberer, *op. cit.*, p. 1.

8 Spiegel–Rösing, *op. cit.*, p. 35.

9 Schmandt, J. (1977) 'Science Policy: One Step Forward, Two Steps Back', in Haberer, J., *op. cit.*, pp. 9–23.

10 Ronayne, J. (1979) *The Allocation of Resources and Development*, Report to Australian Science and Technology Council (unpublished), June, pp. 38–45.

11 Blankenship, L. V. and Lambright, W. H. (1977) 'University Research Centres: A Comparison of NASA and RANN Experiences' (unpublished), p. 185.

12 Mogee, M. E. (1972) 'Public Policy and Organizational Change: The Creation of the RANN Program in the NSF', unpublished MA thesis, George Washington University.

13 Ronayne, *op. cit.*, ref. 10, pp. 1–9.

14 Schmadt, *op. cit.*, p. 14.

15 Spiegel-Rösing, *op. cit.*, p. 29.

16 Senate of Canada (1973) *A Science Policy for Canada*, Vol. 3, Ottawa: Information Canada, pp. 629–643.

17 Brooks, H. (1973) 'Knowledge and Action: The Dilemma of Science Policy in the '70s', *Daedalus* 102, pp. 125–143.

18 (1979) *National Science Policies in Europe and North America*, Paris: UNESCO.

19 (1971) *Framework for Government Research and Development*, London: HMSO Cmnd. 5046.

20 Brickman, R. and Rip, A. (1979) 'Science Policy Advisory Councils in France, the Netherlands and the United States, 1957–77: A Comparative Analysis', *Social Studies of Science* 9 pp. 167–198.

21 Ronayne, *op. cit.*, ref. 10, pp. 1–9.

22 *Ibid.*

23 Price, D. J. de Solla (1969) 'Letter to the Editor', *Science and Public Policy Study Group Newsletter*, November, p. 16.

24 Weinberg, A. M. (1978) 'The Indispensability of Scientific Choice', *Minerva* 16, pp. 339–342.

25 Mr. Gorton made these comments in an interview with B. Nelson. See Nelson, B. (1968) 'Australia: Education and Science Are Looking Up Down Under', *Science* 160, p. 173.

26 Sometimes science policy analysts who are unfamiliar with the Australian system can be misled into thinking that there are exact parallels with other systems overseas. Wilson, for example [(1979) 'Science Policy Institutions: Some Canadian and Australian Parallels', *SCITEC Bulletin* 9, pp. 11–18], equated the Australian Science and Technology Council with the Science Council of Canada and the Australian Department of Science and Technology with the Canadian Ministry of State for Science and Technology. These are very misleading parallels.

[27] Senate Standing Committee on Science and the Environment (1979) *Report on Industrial Research and Development in Australia*, Canberra: Australian Government Publishing Service, pp. 83–85.

[28] Ronayne, J. (1978) 'Scientific Research, Science Policy and Social Studies of Science in Australia', *Social Studies of Science* 8, p. 361, contains an extensive bibliography. See also Rubenstein, C. L. (1978) 'Changes in Australian Science and Technology Policies: From Ends to Means', *Australian Journal of Public Administration* 37 pp. 233–256.

[29] Australian Labor Party (1967) *Platform, Constitution and Rules*, Canberra: Federal Secretariat.

[30] Whitlam, E. G. (1970) 'A National Science Policy', *Search* 1 p. 135.

[31] (1974) *OECD Examiners' Report on Science and Technology in Australia*, Canberra: Australian Government Publishing Service.

[32] (1975) *Science and Technology in the Service of Society – The Framework for Australian Government Planning*, Canberra: Australian Government Publishing Service.

[33] Fraser, M. (1974) *Hansard Report*, House of Representatives, March 21.

[34] (1975) *Towards Diversity and Adaptability: Report to the Royal Commission on Australian Government Administration by its Science Task Force*, Canberra: Australian Government Publishing Service.

[35] *Towards Diversity and Adaptability, op. cit.*, pp. 18–19.

[36] Gummett, P. J. and Price, G. (1977) 'An Approach to the Central Planning of British Science', *Minerva* 15 pp. 119–143.

HUGH STRETTON

SOCIAL SCIENCE: EDUCATION AS SOCIAL PERSUASION

The theme of this paper is that there has been a widespread change of mind about the role of values in the study of society, but that the change has not had much of the effect it ought to have had on the curriculum and teaching of the social sciences.

To compress that theme into the time available entails some drastic simplifying and generalizing. I acknowledge in advance that there have been many individual exceptions and variations to the summary observations that follow.

A PROFESSIONAL MISTAKE

In the recent history of the social sciences, especially in the English-speaking countries, I think there was a massive aberration — a scientific mistake reinforced by the professional and class interests of many of the scientists — from which we are only now making some recovery. The most general name for the phenomenon was perhaps 'positivism'; a common feature of its many forms was a belief that there could be a value-free science of society, and therefore a neutral, non-political role for social scientists 'as scientists', whatever they might value or vote for as citizens.

Obviously, 'positivism' lumps together a wide range of people and ideas. For example there were behaviourists who hoped to understand human society much as ecologists understand rabbit society, from physical observations alone without reference to the humans' ideas or conscious purposes. At another extreme were philosophical idealists who thought that society *was* its structure of meaningful roles and rules, to be understood much as a language is learned and understood — but some of whom still believed that such meanings could be studied in an objective, value-free way.

The common thread of those diverse beliefs ran like this: There is a fundamental difference, characterized most famously by David Hume, between 'is' and 'ought'. 'Is' stands for objective knowledge of the world, 'ought' stands for values derived from passions and articles of faith, and neither can be derived from the other. Plenty of people accept the distinction itself; positivists are those who take a further step — or logical leap — to assert

123

R. W. Home (ed.), Science Under Scrutiny, 123–137.
© *1983 by D. Reidel Publishing Company.*

that scientists should only say 'is', never 'ought', and should never let their values affect their understanding of the world. In particular, social scientists can and should describe society as it is; if asked to do so, they should forecast the likely outcomes of alternative policy choices; but they should leave all the valuing and choosing, and the conflicts of faith and interest, to citizens and politicians.

Positivists do allow a limited role for values in social science. They believe scientists should value truth. They concede that values must help to choose questions for research and to allocate research resources. They also concede that scientists are human and prone to bias, so that a lot of social science is in fact 'value-impregnated' — but that, they say, is a human failing, to be fought by stricter scientific discipline.

During the last decade there has been a widespread shift of opinion away from that positivist faith, for scientific as well as political reasons — if actually applied in practice the positivist faith would make social science technically incoherent, as well as socially irresponsible. But in one version or another, the faith was taught to a great many students in a great many undergraduate and graduate schools of social science through the nineteen forties and fifties and sixties. (For representative samples, see the opening of any edition of Samuelson's *Economics* — there is still not much shift of opinion among orthodox economists.) Those students have grown up to be a fair proportion of today's owners, managers and expert advisers in business and government, and a more-than-fair proportion of today's social science professors. If the faith was mistaken, what may be the scientific and social effects of the mistake?

POSITIVISM AND THE DISCIPLINES

If a value-free social science is not even logically or ideally possible, efforts to achieve it will not succeed, but they can have other effects. The desire to make the social sciences more like physics tended to produce formal, super-ficial or merely aesthetic imitations: sociology *looked* more like physics, without functioning a bit like physics. Theorists tried for high abstraction, for very general laws, for simple motive forces (wasn't material self-interest like gravity or magnetism?) from which complicated processes could be deduced. They tried to deal only in identities which could be objectively measured, and relationships which could be mathematically modelled. And so on. Books could be written (and a number have been) about the detailed damage which these inappropriate aspirations did to particular disciplines. Here a very few examples will have to do.

Sociologists who were possessed by the new ambitions tended to lose interest in actual societies and their problems. They stopped studying communities, institutions, families, or problems of poverty or unemployment or suburban anomie or ghetto conflicts. Instead they began to look for formal general theories of social cohesion or development. In the nature of things those were hard to find, so increasingly they thought and wrote and taught introspectively about the problems of theory and method within their discipline. Fewer and fewer journal articles reported discoveries about society; more and more reported trials of this or that method. Because the new scientific ambitions were alluring but intrinsically sterile, what they actually produced was a shift of interest from the society to the science: I doubt if any discipline ever spent as much of its time introspecting its own frustrations as American sociology did through the nineteen-fifties.

An early source of 'hard-nosed' behaviourism was the University of Chicago. From the nineteen-twenties, its urban sociologists and geographers set out to be 'urban ecologists' and to develop simple theories which would fit and explain the geographical pattern of growth of all modern cities. Actual city life and growth do not have the sorts of uniformities required for that sort of theory, so the attempt was sterile. It nevertheless persisted for a very long time, because the desire for a particular type of scientific theory had displaced any real desire to understand, let alone improve, the life and growth of the cities.

After the war Harvard outdid Chicago as a centre of scientistic ambitions, which became quite Newtonian as Talcott Parsons' followers dreamed of a master-theory so abstract and all-embracing that all other social theories (and disciplines) would become components of it. The dream was sterile, but, for a time, extraordinarily successful in displacing other concerns in the curriculum, the journals, and a great many academic appointments. There was a steady fall (since measured) in the proportion of journal articles about real social problems, conflicts, or policies. There happened also to be a direct conservative bias built into the new theoretical aspirations. Radical voices protested here and there, but as a whole American sociology never did as little for American society, or for real science, as in the quarter-century from 1945.

The parallel history of economics is less sterile, more practical, but perhaps even more reactionary. I believe the useful thing for economic theorists to do is to decide which economic effects are important, and then to study how to explain them by reference to those of their causes that seem to explain them best, or allow the most useful manipulation of them. (For example, employment, productivity, and the distribution of income are important: so which of their conditions is it most interesting and useful to know about?) Orthodox neo-classical economists flatly reject that approach. Instead they try hard to deal only in economic variables, and in theories designed to model relations between economic variables. They like to relate economic effects to economic causes rather than to other social or political causes, partly from social conservatism, but partly for the scientific reason that relations between their abstracted economic system and its social context can rarely be modelled very theoretically or mathematically. Thus a professional value judgment — a preference for dealing only in variables of a particular kind — is substituted for the social purposes which would often be more rational guides to theory-choice and explanation. It happens that the internal relations of those preferred economic variables do not suffice to explain much about the distribution of wealth or income, so the science has some systematic blindness to those subjects of radical concern. But that is only the beginning of the damage the professional values do. 'High' theorists would like to deduce the most they can about the economic system from the least and simplest premises. For that purpose it suits them to deal in a single motivation and a universal measure of economic value. In practice that means a narrowly conceived material self-interest or utility maximization as the motivation of economic activity; and it means that the economic system is defined as including only the activities that are (a) selfishly motivated and (b) subjects of public pricing and exchange. Whatever they sometimes say to the contrary, most of them then proceed to teach their students that that narrow selection from the wide range of material productivity and mutual service *is* the economic system, and works most efficienly when it *does* adhere as closely as possible to those self-interested, open-market, exchange-value principles.

It is possible, however technically mistaken, to arrive at that sort of economic science from purely scientific motives. But — for reasons for which there is no time here — that sort of economic science also serves other interests. To define and model the economic system like that, with tests of efficiency like that, and to teach students to think like that, is in a broad way to serve the interests of adults against children, men against women, rich against poor (including white against black), and capital owners against the rest. It is also to recommend the values of selfish, competitive, unequal societies against any efforts to make them more generous, cooperative or equal.

POSITIVISM AND THE STUDENTS

The ill effects on scientific theory and technique and on the use of scientific resources have been bad enough, but the ill effects on students may prove to have been worse, and harder to cure.

Students were taught, authoritatively, that their values were, almost literally, *childish*: beliefs acquired at home or at church, by faith or authority. They arrived at university at the age when people are most vulnerable to suggestions that it is time to leave home, grow up, reject authority. And sure enough their teachers told them that they must leave all their soft, childish values outside the classroom and laboratory, and learn to think for themselves as adults and scientists, not by faith or authority but by observation and reason, objectively.

Values continued to be important, of course, but in life rather than in science: to husbands, lovers, democratic citizens, but not to scientists as scientists. Indeed (they were warned) scientists had a good deal to fear from values, both their own and those of other people. Illiberal authorities might try to censor any fearless truth-telling science, as the Church did in the case of Galileo. And in insidious psychological ways the scientist's own values could bias his observations and censor his reasoning, if he let them. Far from being the rational organisers of social science, values were its enemies, and should as far as possible be expelled from it.

You may think I exaggerate or parody what those generations of students were told. Having read some dozens of text-book introductions and heard some dozens of introductory lectures to courses in economics and sociology and political science through those decades, I do not think I do. Of course there were many lecturers who did not talk like that. But tens of thousands did, a lot still do, and hundreds of thousands of students learned to regard 'value-judgment' as a sneer word. 'But that is a *value-judgment*', uttered in tones of horror or pitying condescension, has terminated a great many scientific discussions at the point where serious discussion ought really to have begun. Worst of all perhaps, because it increases the difficulties which now face attempts to reform either the science or the society, those students got an insistent message to the effect that moral thought was at best a second-rate, less important kind of thought. I do not think it is entirely fanciful to see some effects of that in the amount of hard-faced uncaring class and professional selfishness to be found in academic ranks today.

Extremes breed extremes. When students began to rebel against those orthodoxies, some went to the opposite extreme of 'thinking with the blood'.

The issue was further complicated for any Marxists among the rebels: Marxist social science is built squarely on social values, but Marx always denied this, and, like any positivist, claimed that his analysis of society was value-free.

To return: If the work of social science is not organised rationally by its social purposes, what else can organize it? Simplifying even more than usual, it is possible to sort the social scientists of the positivist decades into three groups. Some avoided the positivist mistake and continued to acknowledge and apply their social values in rational ways. A second, often larger group did much the same but tried to conceal the values and pretend they were not there, or were being eliminated as the science 'matured'. Since the easiest values to conceal are those of the status quo, this 'value-concealed' science was usually conservative — a lot of the applied kind of neo-classical economics was like that. Third, social values can be replaced by various professional or methodological or even aesthetic values. That happens when people try for particular structures or styles of theory, or for desired densities of footnotes or equations on the page, not to improve real understanding of the world, but for their own sakes or to enhance the authors' self-images as scientists. Some of the 'grand theorists' of post-war sociology and political science hoped for theory like Newton's, others hoped for theory like Norbert Wiener's cybernetic theory. Some modern econometricians try to model real economic relationships for useful purposes, but more and more of the 'purer' mathematical economists seem to be choosing their premises and theoretical forms not for any likely discovering power, but for the elegance and intellectual interest of the mathematics they will allow.

Sterility does not necessarily doom a social science, unhappily. Plenty of sterile theorists have founded journals to publish like-minded work — work which elaborates structures of sterile theory, translates old thoughts into new jargon, or wrestles with problems created in and by the sterile structures themselves. The increase of student numbers has greatly increased the staff numbers, and they have all got to publish something, somewhere, somehow, and make libraries buy it and students read it. I need not labour that familiar theme. The growth of professionalism in the social sciences has all too often been accompanied by a decline of self-knowledge and real productivity, as the profession substitutes its own sectional interests and values for the generous ideas of social good which have organised most of its best work in the past.

The social causes and effects of ideas and teaching practices can seldom be traced with any certainty. Did the intellectual mess I have been describing owe most to the professional and scientific ambition of the scientists, or to their class interests? Does society get the social science its owners and managers want, or does the science shape the society to any degree? For myself, I don't think the professional and class interests create the science in any automatic or exclusive way. But those interests certainly dispose some scientists to accept a lot of bad work which happens to accord with their interests, while attacking good work which does not accord with them.

There remain some grounds for hope. If the science never did anything but express the class interest of the class to which social scientists inevitably belong, then it could scarcely include such conflicting ideologies as it does, and it could scarcely have changed its directions as much as it has through a century in which the class affiliations of the scientists have not changed much. But the social sciences *are* diverse and conflict-ridden, their directions *do* change substantially from time to time, and the radicals among them *do* sway majorities from time to time, in quite important ways. There have been times and places when students were taught that the social sciences existed to improve the justice and humanity as well as the efficiency of social arrangements. Some of those students grew up to contribute to the research, theory, persuasion, and legal and institutional design which produced a good many of whatever decencies our societies possess, from democracy and progressive taxation and general education through old age pensions to the modern structures of social security and (for a while) free health care and full employment.

So I believe there is some scope and some hope for reform by intellectuals like us. But I also believe that some at least of the current malfunctioning of our social and economic institutions, and some of the reactionary uncaring spirit of our public economic policies and private tax revolts, are due to the widespread intellectual corruption of the social sciences, especially in their educational role, brought about by a generation of intellectuals like us.

WHAT SHOULD WE DO INSTEAD?

Discussion must start with the role of the values and ideologies that are being wrongly left out of the curriculum: to put them back in, we need to know what they are and what they do. This is not the occasion to go very deeply into the moral philosophy or the scientific methodology of the subject. Merely to remind you of the outlines of the debate, and to introduce discussion about restoring the debate to the curriculum, I offer three samples: three ways of describing the irreducible normative elements in social theory and research. The first is from Julius Kovesi, *Moral Notions* (Routledge and

Kegan Paul, 1967); the second is home-made for this occasion; the third is from Martin Rein, *Social Science and Public Policy* (Penguin, 1976).

Kovesi disagrees with the fact/value dichotomy itself. He does not believe that language and its meanings can be distributed into two boxes labelled 'descriptive' and 'prescriptive', or 'is' and 'ought'. He starts instead from a different distinction: Aristotle's distinction between material and form. You can describe the materials of which a table is made, and the way bits stick out at its corners, and as much more physical description as you like, but you will not succeed in defining 'table' or telling me what a table is, unless you also refer to what we have tables for, and the functions they perform for us. Most of our language has this character: the words name forms or formal qualities, which make sense only in relation to some human use or purpose. Definitions accordingly have elements of standard or norm — to be a table, the thing must perform its function. Tables ought to stand up insofar as that is one of the characteristics that define them as tables.

So there is no *logical* difference between a formal concept like table and a formal concept like murder. To be a table, the thing has to perform certain functions. To be a murder, the act has to be a wrongful killing. It makes no sense to say 'murder is wrong', because if it was not wrong it would not be murder: 'wrong' is part of the description.

Although there is no logical difference between ideas like table and ideas like murder, there are other differences. They arise from the different purposes for which we have the concepts, and the things they stand for. We have tables to keep food off our laps, we have the concept of murder to prevent types of killing of which we disapprove. We have material purposes, and we have moral purposes. Moral purposes generate moral words or ideas. There are complete and incomplete moral notions. If you hear of a killing, the moral information is incomplete: you do not know whether it was right or wrong until some adjectives are added. But if you hear of a murder you know it was wrong, because 'wrong' is part of the definition of murder, part of the knowledge that it was a murder that happened. Murder is a 'complete' moral notion.

It follows that there is no possibility of separating descriptions of the world from moral judgments of it. All descriptions are for purposes, and though some may sometimes serve multiple purposes, most cannot. If your purpose is moral, and you want moral understanding or judgment, you need a description in moral terms, from a moral point of view. If the description is selected and designed for any other purpose there is no guarantee — and often, no likelihood — that it will tell a moralist what he needs to know in order to arrive at moral judgment of the thing described. So it is rarely possible to make one all-purpose value-neutral description of an act or an institution or a social process, or anything else, such that economists can use it for economic forecasting, criminologists can use it to plan better crime-prevention, politicians can use it to forecast elections, and moral philosophers can use it to 'apply values' and judge how good or evil the reported behaviour is. For those four purposes you may need four different descriptions, each designed with its particular purpose in mind. You may need more still if you need to satisfy several economic forecasters who happen to have different values. In Kovesi's shortest summary, moral statements differ from other statements to the extent that

the recognitors which define words are selected not from our desire to know or manipulate, but from desires to promote or avoid (p. 53).

It seems to me to follow from this that it would rarely be possible to separate moral from other thoughts, even analytically, in the activity of social science. People in society and people in social science usually act for more than one reason. For example they commonly want to *know* in order to be able to *manipulate* in order to *promote* or *avoid* something. For that chain of purposes they need descriptions of the world from a simultaneously moral and practical point of view; and that cannot really be reconciled with the positivist view that 'cognitive' and 'normative' thought can not only be separated, but can be supplied by different people, for example by social scientists and politicians respectively. It is rarely practicable to have the scientists do all the discovering but none of the judging, while other people (politicians, citizens) accept value-free knowledge from the scientists and then, on the basis of that knowledge, apply values and choose policies. If the science is to be of any use for purposes of judging and choosing, it needs to be designed for those purposes from the beginning.

For myself, I find Kovesi's approach illuminating and believe we would understand both science and social life better if we understood them in his terms. But it may still be easier for many people, including me, to debate the issues in the language we are accustomed to, which is the conventional language of the is/ought or fact/value dichotomy. So for those who find that language easier to think in, here follows my second sample, a way of describing the role of values in social science using the conventional language of the positivists themselves.

Social life is very complicated. For reasons there is no time to go into here, it may include elements that are intrinsically unpredictable. (One view of social science and policy research describes them as studying how best to behave in conditions of unmanageable complexity: how to choose and how to apply minimax or least-risk principles in those conditions, for example.) Whatever the nature of the complexity; investigators have to put up with knowing less than everything about it. They have to choose what questions to ask, and what general types of answers to accept and act on. Both choices have to be guided partly by the objective facts and partly by the investigator's purposes in wanting to investigate society. Those latter are social purposes of one sort or another, often controversial and always value-based.

That much, most positivists admit. But the investigator's social purposes not only choose the question, they frequently have to guide the work throughout. Here are three general kinds of reasons why they have to do that:

(1) *Some effects of human consciousness and will*: I leave aside the philosophical problems of will and determinism. There are other problems just as troublesome. First, there are the notorious Hawthorne effects and self-verifying and self-falsifying effects. If you observe an object accurately, it may *for that reason* change itself to falsify your observations; the more people learn about their social behaviour, the more they may tend to change it; the more surely social futures are predicted, the more effectively people may act to avert them; and so on. Second, there are a great many qualities and quantities of social life which nearly everybody agrees are important matters and central subjects of social science, but which cannot even be recognized, let alone measured or compared or explained, without the aid of value judgments. Nor can those value judgments be avoided by substituting the observed values of the society being studied. (They can't really be made to organize the study of themselves; if they could, it would still be necessary to choose which of them should organize the study, for they are never unanimous or compatible; etc.) The things which cannot be recognized, measured or compared without value judgments as well as factual observations include economic productivity, social freedoms and constraints, distributions of power, equalities and inequalities, feelings of community or integration or alienation, the diversity or creativity of societies, and various other moral, aesthetic and inventive qualities of social life.

(2) *Naming of parts*: The continuous structures and processes of social life have to be broken arbitrarily, analytically, into bits whose relations to one another are then studied. There are innumerable possible ways of fragmenting the whole and naming the parts for purposes of study: of deciding what identities to use, and when and why and how to aggregate or disaggregate

them. The social purposes of the investigation offer the *only* rational guide to the choice of concepts, analytical frames, and so on. Unless they are governed by those purposes, any other bases of choice — for example, methodological preferences for certain over uncertain knowledge, for quantities over qualities, for more general over less general categories — are simply irrational, however fashionable some of them may have been at times.

(3) *Endless chains, seamless webs*: Social relations are continuous: chains of cause and effect run from past through present to future, and there are complicated webs of inter-relation at any present time. So what sort of explanations do you want? How far (with diminishing certainty) do you want to predict ahead? How far back into the past do you want to trace the networks of cause and effect? Which ranges of present social relationships do you want to explore? When tracing these networks there can be no purely technical or objective way of deciding in which directions to go, or how far to go, or where to stop. The social purposes of the investigation are the only rational guide to these scientific choices.

None of this reduces the need for research to be objective wherever objectivity is appropriate. Social scientists should certainly observe and measure and count as accurately as the facts allow, and avoid the sort of bias that makes for wishful thinking, careless observation, unrepresentative sampling, unfair selection and the like. But the objective information must always be selected and ordered by theories, conceptual languages and analytical schemes chosen with reference to the social purposes of the research, and the values which those purposes incorporate.

The conclusion of this approach is that if you want to insist on the traditional dichotomy, classifying everything as either fact or value, then you must understand all social knowledge as a closely-woven fabric, with facts and values as warp and woof. If you succeed in persuading the scientist to leave out as much of the woof as he consciously can, in the name of expelling values and reducing bias, that will not make for a better fabric. There will not be any fabric. There will only be a useless heap of unrelated threads of information.

My third sample is here for a double purpose. It is yet another way of describing the role of values in social science. It is also a good text for my main purpose, which is to discuss what the education of social scientists should do about the values in social science. The book is Martin Rein's *Social Science and Public Policy*. I think the title is in one respect misleading. The argument of the book does not apply only to applied or policy-oriented science, as if there is some other, purer social science to which the argument does not apply. It applies squarely to all social science, by reason of the quality that defines it as social.

Rein's account of the role of values in social science is quite elaborate, and I can cite only a small part of it. In that part, he reminds us that people can only see reality — or search it, for purposes of research — through specific frames and selective screens; and those frames, schemes of analysis, conceptual languages and so on are chosen with an eye to the purpose of the search as well as the nature of the thing searched. He instances three familiar examples:

(1) There are social researchers whose 'screens' assume that the causes worth knowing of most contemporary social problems are to be found in the malfunctioning of institutions. Research methods are accordingly designed to find and characterize and where appropriate measure the malfunctions — and such methods will not usually uncover causes of other kinds.

(2) There are researchers, especially in fields related to social work, who assume that most social troubles signify some failure of people to cope; so their methods are designed to find which people are failing to cope in what respects, and why they are failing and what might be done to improve their performance. If the troubles in question happen also to be affected by sunspots or the rise of the middle class, those methods will not detect it.

(3) There are researchers of all political persuasions from Right to Left who believe that to understand social processes or to manipulate them, the main thing to know about is power. They use 'screens' which direct their research accordingly. There may be any amount of relevant cultural or psychological or other information, which those 'screens' will not reveal.

Between those three approaches there are of course questions about which of them is technically appropriate to particular problems. There are also questions about what matters, what is defined as a problem and what might be defined as a solution to it. There are also questions about what sort of knowledge to seek, and these often reflect questions about the kinds of social intervention or non-intervention or self-change that are thought desirable. So the values 'screen' the work throughout, not just at the initial stage of choosing subjects and asking questions. Because they do, there should be continuing attention paid to the screens that are being used; and unless the society and the researchers have identical values and assumptions about the way the world works, there is no reason to expect perfect scientific consensus about the methods in use.

Thus when Rein argues that the values in question should *be* in question throughout the research, he does not merely mean (what positivists would agree with) that the investigator may need to study the values of the people he studies. Rein means that the investigator's own values, and the values built into his research methods, should also be open to question throughout.

What is meant by studying those values, or keeping them in question? Positivists generally agree that scientists may need to know what social effects follow from the values which people believe in; but positivists have generally left it at that, because values seem to them to arise from people's passions or arbitrary will, so that they are not open to reason or useful study or criticism. The values of the people being studied are primary data, the values of the investigators are first premises of the research, and neither is likely to repay further study.

It is at this point that Rein's argument becomes particularly interesting, and educationally practical. He thinks values are by no means opaque to reason. They can be understood, inter-related, confronted and argued about in many fruitful ways. He calls his approach a 'value-critical' one, and here is his own shortest summary of what he means by it:

A value-critical approach subjects goals and values to critical review, that is, values themselves become the object of analysis; they are not merely accepted as a voluntary choice of the will, unamenable to further debate. Most attempts at the analysis of values ask the analyst to discuss alternative means to an end which is accepted without question; or to expose these instrumental means to review by measuring, or taking account in other ways of the short- and long-term consequences of pursuing a particular purpose; or to treat the factual aspects of values and examine their distribution in the population; and/or to study shifts in values held by different groups over time. All these approaches have merit, and I do not reject the positivist tradition in which they are carried out. But I do believe that they are insufficient, for I consider the crucial task in policy review to be the analysis of goals in their own terms, in relation to: (a) the intrinsic meaning of collective values when translated into social purposes; (b) the relationship to other goals with which they may be in conflict; (c) the question of priorities which arises from the pursuit of goals which have equal attractiveness (when they cannot all achieve their maximun value at the same time); and (d) economic and political constraints which must either be accepted or redefined.

In the value-critical approach, not only are values treated as the subject of analysis, but it is assumed that analysis can never be independent of the values we hold. They constitute the framework which helps us organize, not only the problems we address and neglect, but also the inferences we draw. ... Information and data can never be understood in isolation from the context of ideas which give them meaning. And it is these frames, or modes, or values, or ideologies, or theories, or whatever we choose to call them, which are crucial for any creative work... (*Social Science and Public Policy*, pp. 13–14).

Rein suggests that social science should be done in that spirit, and so should education in social science. The values built into (say) an economic theory should be debated as that theory is learned in economics class, not just in some other class in methodology or moral philosophy. I will return to his argument in a moment. First I think there may be a case for adding to it. Students should certainly study and criticize the value-structure of the theories and methods they are learning to use. But it may also help if they have other samples to compare, general themes to reflect on, and examples of explicit political and social philosophy. So there may be a case for some general study of the forms that have been taken at different times and places by varieties of conservative values, liberal values, socialist or egalitarian values, feminist values, puritan values, and so on; and of the social uses or effects such values appear to have had in various historical circumstances. Some schools of social science offer some history of ideas to their students. Perhaps more should do so. To make the most reactionary proposal you are likely to hear at this conference, perhaps wherever there is a compulsory Statistics unit there should be equal time for compulsory Plato, Vico, Rousseau, Mill and Marx.

To return to Rein's argument: a value-structured science applied to a somewhat unstable, self-changing social life cannot produce much cut-and-dried certainty. As to what it can produce, I quote Rein again:

Research within the value-critical framework does try to discern patterns; it seeks general principles that take account of the context and co-mingle facts and values. These general principles cannot be derived solely from the aggregation of known regularities in the behaviour of separate events. Rather, the analyst's advice to policy-makers is based on social understanding and depends upon the use of illustrative stories, or accounts from past experience. . . . The central element in normative story-stelling is the metaphor. This means that we must rely upon actions or events that appear to be analogous to situations we already know and that permit us to reason from the familiar to the un-familiar.

A little later he uses a most suggestive phrase —

The stories we tell can be imaginative, inventive or romantic, using experience in very subtle ways — rearranging it, combining aspects of different experiences. . . (*Social Science and Public Policy*, pp. 74, 76).

That rings true to me — true to the actualities of useful social theory and research. A hundred years ago a few people reflected on a diverse list of things — economic growth and the surplus it was beginning to produce; Pasteur and the agricultural revolution and the rising average age at death; the dearth of charitable services in the new industrial towns; various Christian, humanist and other philosophies of social obligation — and combining bits of all those items, they conceived the idea of a workable old age pension for all who needed it. Some decades later other thinkers drew together some facts of racial difference, some myths of racial quality, some novel morality and some modifications to the technology of the abattoir, and designed the extermination camps. Those two were mild advances of their respective kinds, compared with some of the possibilities that are opening now with our new means of economic productivity — and of genetic engineering, and homicide-at-a-distance.

So I end by offering another metaphor to add to those Rein discusses in his book. As he almost said in the last quotation above, social science is above all a *recombinant* science. When we grasp how like it is to recombinant genetics, we may also begin to grasp what the natural scientists knew about recombinant genetics the moment they conceived it: that its intrinsic technical uncertainties made questions about its management and morality critical for any society that nourished it. In less dramatic ways the social sciences are like that too. I do not believe we should therefore lock them away or censor them. But I do believe that their values and value-structured methods should routinely get their proper half share of teaching time and attention, as they have not done in higher education for a generation past.

University of Adelaide

ROM HARRÉ

HISTORY AND PHILOSOPHY OF SCIENCE IN THE PEDAGOGICAL PROCESS

The argument to come is organized around two issues: the relation of a student at school or in higher education to the research process; and the place of history and philosophy of science in our intellectual economy. I hope to show that a particular way of construing history and philosophy of science (hereafter HPS) provides just the device needed to resolve one of the main problems in the pedagogy of the research process. Further, I hope to show by way of example that it was by the use of HPS conceived as science criticism that it has been possible successfully to revise and so to rescue the central social science of social psychology. I use this example to show how HPS can be used as part of the research process itself.

INTRODUCTION: RESEARCH AND THE PEDAGOGICAL PROCESS

The mistake of a heuristic (or 'Nuffield') approach to learning was to suppose that the intellectual route to a great discovery was so straightforward an inference from evidence, according to some unspecified but relatively simple schema, that it could be recreated by anyone with the same information as its originator. The act of discovery, it was presumed, did not depend on the special knowledge, situation or talents of the discoverer. Experience has now confirmed what common sense foretold — that even with all the necessary evidence before him, a student of average ability is not very likely to remake the great discoveries for himself. It is more than likely that the ways great men of science reached their novel ideas were idiosyncratic. To leave a gap between evidence and discovery, to be o'erleaped by the student in a simulation of the research process, information must be restricted; but without that information most students cannot make the jump. There has, therefore, been a slow return to formal methods in physics teaching, even though a grasp of the research process clearly remains a desideratum in the cultural apprenticeship of citizens who every day depend upon its products, and in the last analysis finance its endeavours.

But what of the student as researcher? Even at the doctoral level, research in the natural sciences is usually no more than the application of routine methods in what we might call the 'atelier' style. (Rubens allows you to

139

R. W. Home (eds.), Science Under Scrutiny, 139–157.
© *1983 by D. Reidel Publishing Company.*

colour in the hand, but you must do it in his style.) But the social sciences
could avoid a merely routine apprenticeship. Even high school projects are
often quite interesting pieces of descriptive ethnography, sometimes better
than professional work influenced by false images of science. Amateurs can
contribute to social science, because to manage their everyday lives they
must have available an everyday competence to understand the social world. I
will be bringing out the consequences of this thought later in the paper.

There is another form of student involvement in the research process,
namely that in which the student appears as a research object. In many universities, so many hours of obedient performance in the subject panel is
a prerequisite for credits in 'Psychology 201'. But in what role does the
student appear? Characteristically, the student as research object is not
encouraged to theorize about the events in which he takes part, or to
comment critically upon them. In one experiment, a student was required
to sit for twenty minutes with his left hand in freezing water while, with his
right hand, he gave another student electric shocks as an encouragement to
learn something quickly. He was not allowed to comment upon this bizarre
event, or if he did his opinions were not recorded. This curious arrangement
was supposed to lead to insights into human aggression, useful, say, to governments in the fight against violent crime; but it is surely hard to imagine a real-
life analogue for it. It depends upon the assumption that the student will
react like an automaton to the treatment to which he is subjected.

It is clear that the physical sciences are too far along to lend themselves
at all naturally to general student participation in the research process as
other than mere apprentices. But this is not true in psychology and the social
sciences. If, as I shall argue, we come to use HPS as science criticism, in ways
parallel to the way literary and music criticism is used in the creative processes
of composition and interpretation, we might in these sciences be able to
realize the Nuffield idea much more fully than has proved possible in the
physical sciences. Further, it might be possible with the help of HPS to devise
a reformed psychology that does involve students, but now as both topic and
resource. Then one might partially meet, at least in this field, the implicit
desideratum behind the Nuffield idea, even when the students are themselves
the research objects.

In recent years, HPS criticism has had a major effect on the human
sciences, particularly psychology. One of the central tenets of the new psychology is an insistence on the fact that ordinary folk are quite competent at
formulating psychological theory. The theories used by ordinary people can
be abstracted from social talk and re-written in the official rhetoric of social

science. Since students are people, it is easy to see what sort of knot could be tied with a well prepared and properly guided social psychology, in which everyone involved was working within the research process. What I propose, then, is that we use HPS itself as a topic, to improve student understanding of the research process in the natural sciences, and as a resource to forge a human science methodology which takes advantage of the student's own social skills. But it is clear that the relation of the student to the research process must be different for the natural and for the social sciences.

PART ONE

HPS as Science Criticism

HPS enjoys an equivocal and uncertain status in most universities, relative both to the sciences and to history. It is generally regarded as something added on to the sciences proper, and in most universities the scientific community can barely conceal its suspicion of, and even contempt for, HPS studies. In this section, I want to develop an idea of Michael Polanyi's that HPS should be as much a part of science as literary and musical criticism is part of literary and musical activity. Just as musicians expect to improve their appreciation of a work by studying it in a music appreciation class, as opposed to listening to it or playing it, so scientists should expect to improve their appreciation of experiments, theories and so on, by the use of HPS.

To identify the elements of the critical process, it might be useful to compare science criticism with other forms of criticism. Superficially, the most obvious analogue is literary criticism, but I would prefer to look to music criticism for a parallel practice, since I shall be emphasizing the distinction between a critical appreciation of form and a critical analysis of content. Faced with a work of music, a critic tries to reveal two distinctive aspects of the work (setting aside commentary on the particular performance and performer). The first aspect is revealed by the analysis of the work to highlight its musical structure, at both macro and micro level. For example, in discussing two divertimenti by J. Scarzer, H. Ulrich says,[1] 'Each contains *four movements: allegro, minuet* and *trio, larghetto* or *adagio,* and *allegro.* . . . These may be among the first works which omit the harpsichord and its function of *realizing the figured bass* . . . . *Sonata form* is present in a rudimentary fashion'. (Italics mine.)

This sort of analytical activity presupposes existing categories of musical elements and the relation (harmony, for instance) that can obtain between

them. Also, in the case of music, there are important transformational principles, such as inversion, which function as the rules for developing variations
by which macrocomponents, such as movements, are brought into relation.

The second aspect is revealed by an interpretative procedure, explicating
layers of meaning. This may involve quite strong assertions as to the content
of a work. There seems to be a three-way negotiation of meaning — com-
poser's intentions, when known (cf. Mahler's gloss on his third symphony as
the ultimate transfiguration of man in nature); critic's interpretations (cf.
Wagner on Beethoven's symphony as a celebration of the essence of the
dance); and audience appreciation. The last is usually treated as subordinate
to and correctable by the claims of the first two parties. This sort of discussion is rich in metaphor. Ulrich in the passage cited above goes on to say
that these pieces are 'ingratiating, noticably homophonic, and altogether
enjoyable'.

For the critic's commentary to be of any use, there must be a critical
community, defined as those who share assumptions that certain structures
are ideal forms, and that certain meanings are available to the intelligent
listener. An aesthetic of musical form would, of course, need arguments for
the ideality of the forms that are functioning as standards of judgement and
critical practice. And since there is nothing to which to refer aesthetic matters
beyond the conventions of taste, these arguments would usually terminate in
historically conditioned conventions. Music criticism, then, is partly music
history. The same temporal relativity is as apparent in the changing interpretation of works in different epochs as it is in the changing conceptions of
what are ideal musical forms.

Science criticism, too, could, I think, be taken to encompass both the
revelation of the structure of patterns of scientific thought (form) and
critical interpretation of what is actually asserted in those thoughts (content).
Corresponding to the music critic's discussion of particular performances,
there is a critical process in science, culminating in doctoral examinations and
refereeing by editorial boards.

The development of an adequate scheme for analyzing the products of
scientific activity, and the attempt to justify taking certain ubiquitous forms
as ideal, together constitute, of course, a philosophy of science. In recent
years, philosophical analysis of science has been dominated by the assumption
that major analytical tools should be logical, but the structure to which I
shall be drawing attention in my detailed exposition is not the logical structure
of the discourse, but a structure to be discerned within a science, the description of only part of which is scientific discourse. Considered abstractly, it is

the structure of the content of a discourse, both explicit and implicit, and not its logical form, to which I shall be primarily drawing attention.

But why choose to analyze the structure of the content, rather than the more readily identified multiple structure of the discourse? The major short-coming of discourse analysis, as a guide to ideal forms of cognitive structure, is that explicit scientific discourse does not record (because it does not need to record) all the material germane to the formation of the cognitive basis of the scientific treatment of the field. The analytical scheme which I shall be expounding includes reference to the real, generative mechanisms which actually produce orderly, natural phenomena. These are usually unknown in the initial stages of theory construction. Yet the fact of their being unknown exerts a powerful influence on the assembling of cognitive material. As unknowns, they are not in general represented in scientific discourse, since there is nothing positive to say about them. References to the real causes of phenomena would appear explicitly only in a metascientific statement of the aims of science — for example, the assertion that science was an attempt to uncover the generators of natural patterns. Basic background assumptions about the nature of things, too, are rarely described in scientific discourse, because an author and his community of readers usually share them as un-considered cognitive elements. Yet if we are to understand the constraints on theory construction, that is, on the imagining of possible generative mechanisms that could produce observed patterns, this tacit material must be entered explicitly in our account, and its relation to the content of the ex-plicit theory examined.

Another weakness of merely analyzing logically the meagre discourse that scientists provide as a public presentation of their cognitive structures, concerns the representation of substantive content. Internal relations between concepts, reflected in so-called subcategorial rules of language and further reflected in our uneasy feelings of impropriety when presented with such combinations as 'cold fire', are not detected with the standard logical tools. If we are restricted to the predicate calculus as our analytical apparatus, there is no way of distinguishing the semantic complexity of metaphorical usages from the simple referential meaningfulness of more literal readings of language. The use of metaphor is one of the main instruments of scientific advance. Bringing out the non-logical relations between predicates is the interpretative phase of science criticism.

The Elements of Science Criticism

(a) Structural Requirements for a Science. Philosophical analysis suggests that a science requires two conceptual schemes. There must be an analytical scheme for revealing the texture of the world. (We note the possibility that any particular application of an analytical scheme may create merely an illusion of texture.) This analytical scheme provides a classification of natural beings at the level of nominal essences, that is, the level of criteria for bringing material beings under a concept by reference to their observable properties.

An explanatory scheme is required for generating explanations of both the existence and nature, for example the structural properties, of the non-random entities, both synchronic (that is, existing at a moment in time) and diachronic (that is, appearing through time), revealed in the texture of the world. The explanatory scheme provides a classificatory system at the level of real essences, that is, its criteria for class membership refer to the deep and fundamental natures of natural beings, those natures which are not usually directly observable.

To create a taxonomy, a scientific classification system, scientists have characteristically attempted to eliminate the inherent possibilities of arbitrariness or mere pragmatic advantage in classification systems, by using hypotheses about real essences to support a practice based upon nominal essences. Thus a classification of organisms based on morphological distinctions is underpinned by a phylogenetic scheme on the presumption of some general principle of descent, and the joint morphological/phylogenetic scheme is itself grounded in genetics. In the end, the nominal essence species, developed for practical purposes at hand, are supported by hypotheses concerning distinctive genetic codings (qualified by the possibilities of epigenetic effects). This pattern is repeated in most sciences. Reactive dispositions and equivalent weights as the basis of chemical classification are modulated into a nominal essence/real essence taxonomy by use of the electron-structure theory of atomic composition, to explain the observable differences in chemical behaviour.

By drawing on the fashionable but scientifically suspect rhetoric of the bilaterality of the brain and the bicamerality of the mind, we can say that the structural conditions for an analytical and explanatory scheme can be discussed in left hemisphere terms as the properties of a linguistic system determining discourse, and in right hemisphere terms as the organizational principles of a system of analogues or models (that is, concretely represented structures) from which a partial representation of unobserved realities is generated as explicit theory. Such a system would represent the internal content of the theory as expressed in language, together with the tacit component comprising the community's shared assumptions, which might never have been explicitly formulated. Extensional content of the conceptual system arises in the last phase of all, namely in the conscious application of the theory to the world as experienced, when efforts are made to secure the reference of theoretical terms by empirical research. Since I want to talk about content, and particularly the genesis of meaning in advance of experience, I am going to talk in bicameral terms. So if you would care to tiptoe with me across the corpus collosum, we can enter the mysterious and forbidden world of the right hemisphere, inhabited by shifting and incomplete parallels and analogies, where new conceptual relations are forged to emerge in the metaphors which mark the advance of scientific theory.

I call the scheme which represents the structure of the intensional content of a theory or cluster of similar theories the '3-model-model'. It can be set out as follows:

Analytical Model	Patterns in Experience	Generators of Patterns	Explanatory Model	Source Models
→ application	← causation	← simulation	← abstraction	

Patterns in experience are revealed by the application of concepts and perceptual categories derived from an analytical model to a world of experience already structured by commonsense categories. In general, the real but unknown generators of discerned patterns are not empirically observable, so explanatory models, that is, models of unknown causal mechanisms, must be invented to stand in for them. Source models represent the community's shared assumptions, which control the conception of explanatory models; there is much historical, social and metaphysical material that controls the selection and content of source models.

This structure can be illustrated from Darwin's account of his theory, in the introduction and opening chapters of *The Origin of Species*.[2] It is, in fact, misleading to speak of Darwin's contribution as a theory at all: it is really a conceptual system for constructing theories to explain particular episodes in the natural world. But its mode of construction has just the form of a 3-model-model. The analytical model is the *family*, attracting analytical attention to lines of descent and heritable similarities and differences between generations. The source model is *domestic variation and selection,* and the explanatory model is drawn from that by abstraction as *natural variation and selection*. In left-hemispherical terms, each model engenders and controls a conceptual system, linguistically realizable in a scheme of related terms for identifying and classifying the items in a certain field of phenomena and for formulating theories to explain their nature, order and relations.

As I have argued, a complete taxonomy is created by supporting a scheme of superficial natural kinds, carved up from observable likenesses and differences among the items of interest, by making propositions reporting empirical regularities into criteria or rules for applying terms about kinds. Support for this transformation comes from a deep-seated scheme of underlying natural kinds, based on hypothetical distinctions within the explanatory discourse. These are the hypothetical real essences of natural kinds.[3] They are linked to observable differences by the principle that some surface distinctions are reflections of underlying differences; for instance, the deep underlying differences between the proportions and organization of the electrons, protons and neutrons that are believed to be the constituents of atomic types, are reflected in observable differences in atomic weights and chemical properties. In terms of the 3-model-model, the linkage between nominal and real essence that creates the taxonomy of the field is some form of co-ordination between the analytical and source models, tacit acceptance of which determines the content of the relevant theories.

Must the analytical and explanatory models be metaphysically co-ordinate — that is, must they draw on the same source model? That would be the strongest form of co-ordination. The theory of organic evolution seems to be an example of this kind. The history of science suggests that there is a strong pressure towards metaphysical unity or co-ordination of analytical and explanatory models. Sometimes, it is when that unity begins to dissolve that a feeling of crisis develops in a field. The mid-nineteenth-century crisis in chemistry can be looked at in this way. Inorganic chemistry was proceeding by the use of gravimetric analysis, nicely co-ordinate with a simple mechanistic source model of atomic aggregates or clusters for constituent molecules. It became clear, however, that the required differences between organic molecules as the constituents of chemically distinct substances could not be molecular formations based on mere atomic aggregation. Introducing the idea of structured assemblages of atoms allowed organic chemists to proceed with their researches, but the restoration of full co-ordination between analytical and source models became very difficult, and was not fully realised, in fact, until the advent of X-ray spectroscopy.

(b) *The Critical Analysis of Content.* In discussing the kinds of critical activity to be found in music, I distinguished between critical discussion of form and critical discussion of content. In setting out the 3-model-model, I have been doing critical work on the form of scientific thought and the structure of its modes of theory formation. But Darwin's work on the philosophy of his theory includes science criticism of substantive concepts of biology. In the introduction to the *Origin,* a philosophical criticism of the conceptual distinction between a variety and a species breaks the necessary connection that had been maintained until then between two central substantive concepts of biology, species and invariability. It was this conceptual connection which allowed empirical demonstrations of variability, such as those by Nägeli, to be dismissed as having to do only with the establishment of variety. By arguing against a necessary connection between the concept of species and that of invariability, Darwin prepared the way for a doctrine of the 'origin of species' in transformations of prior populations. Conceptually speaking, such a doctrine would have been strictly contradictory, and hence incoherent, if couched in terms of the unrevised traditional concepts of species, variety and variation. If a species was defined as an invariable population group, the idea of such a group having an origin in the transformation of a prior population could not be coherently formulated.

Does Philosophical Analysis, that is HPS as Science Criticism, Reveal Ideal Forms?

Traditionally, arguments for the ideality of cognitive structures have been epistemological. In terms of the 3-model-model such arguments would need to demonstrate that, say, some concept of plausibility could be defined in terms of that structure, so that plausible explanatory models were more likely than implausible ones to be true representations of real causal generators. I believe that such arguments can be produced. They have the form of a demonstration that adhering to cognitive structures which approximate the ideals allows a scientist more readily to achieve what are taken to be the aims of science. It is an important observation that arguments for the ideality of certain forms, such as Hempel's arguments for the DN model of scientific explanations, cannot be fully appreciated without referring them to an explicitly formulated aim of science. Objections to Hempel have turned not only on the implausibility of his scheme relative to actual scientific practice, but also on the extent to which it incorporates, as an unexamined assumption, positivistic views as to the aims of science.

But recently, new arguments about ideality have appeared in the sociological study of science. It has been argued by some sociologists that explanations can be given for the ideality of certain forms in sociological rather than epistemological terms. Both Knorr[4] and Latour and Woolgar[5] have suggested that scientific activities, including adherence to certain ideal forms, can be explained wholly in terms of an anthropology of the laboratory, conceived in terms of revealing the social forces engendered by a capitalist mode of production of artefacts, namely scientific writings. In these terms the ideal forms are to be understood in terms of the need to further members' class interests. Knorr, for instance, has argued that logicality is valued in written papers as part of a defensive strategy for making points in debates in which reputations are at risk. If this thesis were to be successfully maintained — in effect, that all epistemological concepts could be systematically replaced by sociological concepts, without any loss in the power to explain the activities of scientists — then arguments for ideality would be directed to showing that the adoption of a certain form was particularly potent in furthering the interests of scientists in their entrepreneurial enterprises of accumulating symbolic capital, in the form of reputation and respect.

However, still more recent work in the sociology of science, particularly by Collins,[6] has drawn attention to the importance of cognitive interests in the competitive struggle for reputation and recognition in the scientific community. So far as I can see, these comprise the traditional epistemological ends of truth and knowledge. There is a certain irony in this development, since most sociologists of knowledge of the more extreme persuasion have assumed an instrumentalist or positivistic conception of science, while the appearance of cognitive interests suggests the adoption of a kind of realism by most working scientists.[7,8]

The plausibility of much that the sociologists of science have had to say about scientific practice suggests that philosophers should take a middle position in arguments for ideality, one which ought to involve the exploration of both sociological and epistemological considerations. But if we admit to the need to use both epistemological and sociological elements in explaining the acceptance of some form as ideal, then a new problem presents itself. What are the criteria for good explanations of this mixed kind? I do not see how this question can be answered until we, the community, have had more experience of accepting and rejecting instances of such explanations.

But the problem is rather more complex than I have suggested so far. If we follow the suggestion that HPS should be construed as a form of science criticism, how is it related to the moral and political criticism of science? Let us, without prejudice, call the criteria of science criticism, standards intrinsic to an intellectual enterprise, and those of moral criticism, standards extrinsic to that enterprise.[9] There is an obvious and simple argument to show, at least for the physical sciences, that these criteria are effectively independent. If the same physical theory can be held in principle by any society, but the moral consequences of a theory emerge only as it is involved in a programme which is embedded in a particular society, and since moral criteria are culturally dependent, moral judgements of science are culture-relative in a way that judgements of science criticism are not. This would be so whether or not one subscribes to an argument like that of Habermas[10] concerning the social consequences of modelling social and political judgements upon positivistic readings of science. I do not mean to imply that the judgements of science criticism do not have their own mode of culture relativity, but it is, I think, independent of moral relativity.

The situation is rather different in the social sciences. Science criticism of psychology tends to reveal images of man which quite patently have moral and political dimensions.[11] The controversy between behaviourists and ethogenists is partly a difference over analytical models, but more deeply it is a difference concerning the actual and possible autonomy of man.[12] So it is a moral and political dispute. Psychology is, as Shotter has put it, a moral science.[13]

Conclusion: HPS in the Pedagogic Process

If HPS as science criticism can reveal so much, one might be tempted to argue that the problem of introducing a student to the research process is solved: teach him HPS as well as science. But I fancy that extracting HPS as an independent study, separated in the curriculum, will not help to bring about the desired pedagogical result. HPS taught as a topic is not easily transformed into a resource. Nevertheless, it seems clear from the argument so far that HPS studies as science criticism do isolate and identify just those features of the research process that are needed pedagogically. In particular, they serve to identify ideal forms, and to make explicit what is tacit. Where, then, ought HPS to be taught as a discipline? The argument leads inexorably to the conclusion: where teachers are taught, that is, in universities and above all in teacher training colleges.

PART TWO

HPS and the Genesis of New Sciences

The arguments of Part One suggest a normative conclusion: HPS can be used to reveal the features that are necessary for a study to belong in the family of established sciences able to deliver classifications and explanations to which we can accord the status of knowledge. It seems natural to ask whether HPS could serve in the genesis of a science. By describing how HPS considerations were crucial in the emergence of ethogenic psychology, I shall try to provide a general defence of the indispensable role of HPS in creating any science. Though the example I shall consider is contemporary, I do not think that this is a new phenomenon. I would equally want to argue that part of Darwin's originality lay in the fact that, as faithfully retold in the *Origin,* he deployed a succession of essentially philosophical arguments and considerations to reach his mature theory. It is instructive to realise that Darwin did not reach his position inductively. His many observations and some few experiments seem to me to serve merely as the basis of anecdotes illustrative of the power of his conceptual scheme, which was arrived at by ratiocination, by the critical contemplation of analogies and by well directed conceptual analyses. Facts, for Darwin, were not to be treated as evidence in the way they are on the inductive model.

Let us turn to the case of ethogenics:

(a) *The Critical Phase.* There is, as yet, no institution of science-criticism, as there is of literary, dramatic or musical criticism. Science-criticism seems to begin *ad hoc* with some sort of crisis.

The critical phase in social psychology began with the convergence of three sources of dissatisfaction:

(i) the feeling that the empirical methods not only were yielding mere trivialities (or artefacts) but that there was something absurd (literally risible) about them, as in the social facilitation experiments of R. B. Zajoncs.[14]

(ii) the appearance of a multiplicity of mini-theories, e.g. cognitive dissonance, balance theories and attribution theory, united only in seeming to be reflections of middle-class American social practices and unexamined social mores. Each theory seemed to reflect a local cultural stereotype of 'proper' behaviour, rather than any deep or universal conception of man-in-society.

(iii) the growing realization that there was a source-model behind the whole system of institutionalized practices that had been presented as an objective science of social psychology, one which was highly restrictive and extremely implausible. This source-model was merely a generalization of simplistic learning theory, and was thus based on the unexamined assumption that all forms of human action could be decomposed into simple paired correlations of input and output elements, institutionalized in a rhetoric of 'dependent and independent variables' and in the practices of peculiar social events called 'experiments'.

Critics drew attention to tacit assumptions, which in terms of the 3-model-model one could see as raising the question of what were the hidden analytical and source models. In the standard presentations of social psychological theories, only explanatory models (hypothetical causal mechanisms) were described explicitly. For example, social exchange theories did not formulate explicitly the capitalist economic source model upon which depended our ability to understand the concepts and the assumed conceptual links in the theorizing. Once the buried models were brought out, the question 'Are all men and women entrepreneurial in their social relations?' could be asked. And, I'm afraid, rather shortly and simply answered. We have a good example of the genre in recent research on 'love' carried out at the University of Wisconsin. According to the researchers involved, the 'love' relation is maintained only if the running cost-benefit analysis for each person turns out positive. What, I wonder, would Eleanor of Aquitaine have made of this idea? A simple 'economic man' model is controlling the theorizing.

The failure to specify an analytical model was more obvious. This represented a deliberate effort to desocialize the interactions being studied by setting them up in anomic, bland environments, in order to avoid 'contamination' of the variables. In these circumstances the question of how far forms of social action are reciprocally related to scenes, settings and situations could not be raised. Nevertheless, the metascientific or science-criticism question, 'What is your analytical model?' was asked from time to time. It produced such astounding replies as 'We do not have one' and 'Analytical concepts will somehow emerge from pure experience'.

Since 'No perception without conception' is perhaps the one rock-bottom truth of epistemology, it was not surprising to find that those who made such stern replies depended, in fact, on unexamined commonsense categories under new and fancy names. These categories were clearly controlling the analysis which generated the objects of study, that is, which enabled a worker to identify units of social and interactive behaviour. For example, the pseudo-technical term 'risky shift' lumps together a number of distinctive forms of collective behaviour such as 'showing off', 'comfort in numbers', etc., while effectively preventing one from addressing the question of whether to accept the subsumed commonsense categories as satisfactory ways of identifying distinctive kinds of social action.

Revealing a source model and recognizing it as a local cultural phenomenon allows a second phase of criticism, introducing moral and political criteria. The first phase is essentially logico-philosophical, since the error involved in basing an alleged human science on a tribal stereotype is that of concluding from the particular to the general, in the absence of a theory to support such a move. The issue in the moral criticism of psychology turns on the point of human potentiality. There is historical evidence for the principle that an authoritatively promulgated psychological theory can be absorbed by the folk as an ideal-type to which they strive to conform. In short, psychologies tend to make themselves true by bringing about changes in human cognitive function, social relations, child-rearing practices and so on, among those who believe them to be true. By tacitly claiming that people are automata emitting responses stablized by the conditioning of operants, human potentiality is actually reduced by public and influential proclamations by psychologists of 'discoveries' and 'methods' which become true upon the acceptance of the claim. The second phase of science-criticism in the human sciences involved the moral and political criticism of source models, and since, in a highly ordered science, analytical and source-models are co-ordinate, analytical models were criticized as well.[15]

B. F. Skinner's courageous publication of his highly objectionable moral and political beliefs in *Beyond Freedom and Dignity*[16] brought on a bout of ethical criticism of experimental psychology, just as the reductive and demeaning character of the scientism involved was being pointed to by Maria Jehoda[17] and independently and somewhat arcanely by Habermas.[18] The central intuition amounted to this: that the method of experimental enquiry requiring the partitioning of human actions into atomic elements to be treated as dependent and independent variables, in the learning theory tradition, demeaned and ultimately reduced human potentiality. The philosophy of science assumed by Habermas was essentially positivistic and Humean. Some of his criticism was therefore confused, since he took the morally objectionable positivistic programme for human studies, which was incorporated in the experimental tradition, as defining the limits of a scientific study of mankind. The adoption of a realist philosophy of science is almost sufficient in itself, however, to right the moral order, since this emphasizes human autonomy by defending the conception of men as agents, actors with an indefinitely open potential for devising new selves and new parts for them to play.

In short, science-criticism, through the application of a conception of science that I have tried to codify in the 3-model-model, revealed that the old psychology depended on a source model, namely that men were like programmed automata behaving in accordance with the principles of entrepreneurial capitalism, and on an analytical model based on local (and in this case primarily American) cultural assumptions. The models were closely linked: assuming that action is mere routine leads quickly to the idea of *producing* adequacy in human affairs by training in prior agreed routines. This easily slips into the general practice of behaviour therapy, in which infractions of the customary and moral order are dealt with by retraining the automaton rather than by struggling with the evil tendencies of the selfish or vicious soul. Both models were highly problematic, not just because of their moral and political consequences, but as foundations for a science. Both represented radical and undefended simplifications of the object of research, man in society. As soon as that research object is examined in action in more complex cultures than university psychology departments, the plausibility of both models vanishes. This point has been made particularly effectively by Moscovici, by comparing European socio-cultural assumptions with those current in the United States among the middle classes in the 1950s and '60s as revealed in the unexamined presuppositions of social psychology.[19]

I have pointed out that using philosophy of science as the tool of science-criticism corresponds to structural or formal analysis as it is found in dramatic or musical criticism. Analytical philosophy is needed to complement the critical process, since by using HPS considerations, unattended components of the cognitive structure of a science are brought to our attention, and we are then able to examine the conceptual relations involved. In the case on which I am focussing, the relation between the structural criticism and conceptual analogies is very clear. Commonsense assumptions require (despite the overlying rhetoric of 'behavioural measures') that observers be able to partition the flow of human social and co-ordinated behaviour into actions, and that the flow was produced by people with the intentions so to act. But when old-style learning theory controls the source models, the model of man involved is that of a programmed automaton who cannot *produce* actions to realize intentions. The performances of programmed automata must be independent of intentions, or, at most, merely correlated with the intentions that such automata are programmed to experience from time to time. But philosophical analysis shows that the capacity to identify actions is dependent on a capacity to attribute intentions to actors, since categorizing actions into kinds depends in part upon identifying them as the realizations of actors' intentions. Just as Maclaurin and Leibniz showed that Newton's mechanics, as an analytical scheme for studying the behaviour of material things, was inconsistent with its given metaphysical basis, so science-criticism shows that traditional social psychology involves a similar incoherence. Newton's mechanics required finite forces, but the underlying matter theory involving rigid bodies and action by contact demanded infinite forces.[20] Commonsense assumptions involved in empirical psychological studies presuppose a conceptual link, namely of kinds of actions with kinds of intentions, that is denied by, indeed expressly contradicted by, the content of the old source model.

(b) *The Constructive Phase.* In the light of these criticisms, could the 3-model-model and the techniques of philosophical analysis be used to create a new beginning? The key elements are the source model, determining the metaphysics, and the analytical model, determining part of the methodology. Ideally they should be co-ordinate. A complete methodology requires the formulation of at least some of the range of possible explanatory models, and the examination of the demands which the assumption of their possible reality places upon technique (cf. the demand for a microscope of a certain resolving power and magnification which the bacterial theory of disease placed upon optical studies). In the philosophically controlled development of the 'new psychology', a powerful analytical model and methodology developed very quickly, far in advance of a methodology capable of giving empirical test to the hypotheses derived from the co-ordinate explanatory models.

The generic source model behind most of the group of 'new' psychologies is that of men as active agents, formulating projects of various sorts and attempting to realize them by conventional means. Such a brief formulation conceals a very complex conceptual structure involving a hierarchy of deep assumptions. The power of this source model is a consequence not only of its anthropomorphic character, but of possibilities within it for generating a variety of complementary explanatory models. Such variety is inherent in the complexity of its structure. For instance, it has fathered both the grammatical analogy — orderly social behaviour is produced in the way orderly speech is produced (Clarke[21]) — and the liturgical analogy — informal but orderly social behaviour is produced in the way formal orderly social behaviour is produced (Goffman[22]).

The moral and political aspects of an active-agent source model involve the conservative observation that most people, in most known communities innocent of experimental psychology, believe themselves to be agents, at least in principle. Not to believe that one is an agent is generally treated as a pathological condition (cf. the bicameral voices of schizophrenics or the cosmic despair of the Ik[23]). But they also involve the radical observation that active agents, capable of formulating and realizing personal and collective projects, can create new social forms and new kinds of cognitive operations, and perhaps come to feel novel emotions, never hitherto experienced.

The co-ordinate generic analytical model for social psychology is the dramaturgical standpoint. It suggests that the initial phase of an investigation is to analyze a fragment of social life as if it were a staged performance. Relative to the old psychology, the most important effect of the adoption of this model has been to draw attention to the conditions for the staging of life's events. This has raised the question of how settings and props come to be endowed with meaning, and how the folk as audience of their own actions come to be able to make the right readings of the components of a social scene. Further, it has led to attempts to study the way individuals adopt distinctive styles in setting their personal stamp upon standard role-performances. All kinds of consequential analytical distinctions follow. But to grasp fully how the 'new' psychology sees itself, and to locate the point of advance represented by the models so far formulated, another philosophical distinction is required.

In the debris of the collapse of the programme of transformational generative linguistics, one whole piece of conceptual masonry seems to have survived intact. This is the distinction between competence and performance as a guiding principle for non-behaviourist psychological research. Treating this distinction as a prescription suggests that a psychology of this or that aspect of human functioning must include a theory of competence, formulating in an orderly fashion the ideal repertoire of knowledge and skill required of a typical member; and also a theory of performance, describing the generative process by which actual members realize their knowledge and skill in their conduct on particular and specified occasions. The structure of ethogenic psychology reflects this demand.

Hypotheses about people's repertoire of social knowledge can be arrived at by the analysis of accounts of members' own interpretations of events, in the justificatory commentary with which human social life is overlaid and continuously corrected. Adopting the dramaturgical standpoint is adopting a stranger's view. To get at the meanings and hence at the potency of events for actors, that standpoint must be supplemented by hypotheses about members' interpretations. Without this control, the range of ambiguity in outsiders' interpretations is far too great. The methodology which combines outsiders' analyses with members' interpretations as revealed in the practice of giving accounts is just a projection, into methodology, of the philosophical doctrine that actions cannot be unambiguously identified without reference to actors' intentions and interactors' interpretations.

I can now take the moral and political critique a stage further. If one cannot complete an analysis of the events in question without members' interpretations, then those who used heretofore to be called, contemptuously, 'subjects', enter the scientific research teams of the new psychology as members in good standing. And this involves something morally and politically potent, namely listening to and taking seriously what the folk have to say about their own actions. Methodologically there is yet another step. One must incorporate items from ordinary language into the technical resources of a proper human science. In consequence, the philosophical analysis which reveals the structure and inner order of the psychological vocabularies in daily use must be a proper part of human psychology as a science.

Psychological studies in the ethogenic mould require the participation of human beings in a fashion quite other than what it has been in traditional forms of investigation. If we are exploring the way human beings actively realize personal and collective projects, and knowingly and intelligently maintain meaning and order through talk, then ordinary folk are necessarily involved in theorizing about and analyzing out their actions, in order to be able to perform and defend themselves when challenged. Student involvement in ethogenic psychological studies cannot then be other than full participation in the research process as an investigator in good standing. Since this is true of all studies which involve the attribution of intentions and the interpretation of meanings, most contemporary forms of human studies[24] will present none of the old difficulties of relating students to the research process in an active way. As fellow human beings, they too have the right to categorize and explain.

Oxford University

NOTES

[1] Ulrich, H. (1948) *Chamber Music,* New York: Columbia University Press, pp. 159–160.
[2] Darwin, C. (1859) *The Origin of Species,* London.
[3] Bhaskar, R. (1978) *A Realist Theory of Science*, Brighton: Harvester Press, 2nd edition.
[4] Knorr, K. (1980) *The Manufacture of Knowledge,* Oxford: Pergamon.

[5] Latour, B. and S. Woolgar (1979) *Laboratory Life*, Los Angeles: Sage.

[6] Collins, H. (1975) 'The Seven Sexes: A Study in the Sociology of a Phenomenon, or the Replication of Experiments in Physics', *Sociology* 9, 205–224.

[7] Barnes, B. (1974) *Scientific Knowledge and Sociological Theory*, London: Routledge and Kegan Paul.

[8] Bloor, D. (1976) *Knowledge and Social Imagery*, London: Routledge and Kegan Paul.

[9] Musgrave's insistence (see this volume) on the radical separation of factual and evaluative statements, particularly with respect to their consequences, can be used to underline the argument; but with a qualification. Evaluative statements, if believed, engender facts, because 'I believe that *p*', as a factual statement, describes a fact, which engenders other facts. This is particularly so when *p* is a statement about values, since that is just the kind of statement people are prone to believe. But, in psychology, evaluative propositions describing source models of man entail evaluative consequences, which then appear as if they were factual statements. Hence the simplicity of Shotter's demonstration: since psychology must adopt some model of man, it is a *moral* science.

[10] Habermas, J. (1978) *Knowledge and Human Interest*, translated J. J. Shapiro, London: Heinemann.

[11] Passmore's category of critical philosophy of science (see this volume) is particularly important with respect to psychology because of the interactive effect of psychological theory on psychological reality, via *représentations sociales*; cf. Moscovici's studies in *La psychoanalyse, son image et son publique*, Presses Universitaires de France, 1974.

[12] Hollis, M. (1977) *Models of Man*, Cambridge: Cambridge University Press.

[13] Shotter, J. (1975) *Images of Man in Psychological Research*, London: Methuen.

[14] Zajoncs, R. B. (1966) *Social Psychology*, Belmont, California: Brooks-Cole, pp. 10–15.

[15] Harré, R. (1979) *Social Being*, Oxford: Blackwell.

[16] Skinner, B. F. (1971) *Beyond Freedom and Dignity*, New York: Knopf.

[17] Recently re-iterated most elegantly in her M. Jahoda, (1980) 'One Model of Man or Many', in *Models of Man*, edited A. J. Chapman and D. M. Jones, London: British Psychological Society.

[18] J. Habermas, *op.cit.*, Note 10.

[19] Moscovici, S. (1972) 'Society and Theory in Social Psychology', in J. Israel and H. Tajfel (eds), *The Context of Social Psychology*, London and New York: Adademic Press.

[20] Harré, R., and E. H. Madden (1980) *Causal Powers*, Oxford: Blackwell, new edition.

[21] Clarke, D. D. (1979) 'The Linguistic Analogy or When is a Speech Act Like a Morpheme?' in G. P. Ginsberg (ed.), *Emerging Strategies in Social Psychological Research*, Chichester and New York: Wiley, Chapter 3.

[22] Goffman, E. (1967) *Interaction Ritual*, Chicago: Aldine.

[23] Turnbull, C. H. (1973) *The Mountain People*, New York: Simon and Schuster.

[24] Brenner, M., P. Marsh and M. Brenner (1978) *Social Contexts of Method*, London: Croom Helm.

RANDALL ALBURY

SCIENCE TEACHING OR SCIENCE PREACHING? CRITICAL REFLECTIONS ON SCHOOL SCIENCE

1. INTRODUCTION

From the very beginning of formal science education in the schools, a fundamental dilemma has plagued the design of science curricula. As David Layton
puts it, in his study of the origins of science teaching in the schools of nineteenth-century Britain, 'The question of content or process, subject matter or
method of inquiry, is a recurring and still unresolved issue in the relatively
short history of science teaching'.[1] In our own century the predominant
approach prior to the 1960s was one which emphasized the content of
science as a body of generally stable knowledge, of 'patterns and structures
that have been found by scientists to be useful ways of ordering the world'.[2]
In the last fifteen years, however, curriculum reform has led to the introduction of 'discovery methods of learning' which attempt to present science as
'a system for helping people process new phenomena or events that may not
be consistent with the patterns and structures they normally use or are not
consistent with the accumulated body of science dogma'.[3]

Both the content approach and the process approach to science teaching
have their advantages and their disadvantages, and I shall briefly characterize
these in the section which follows. Then I shall raise the question whether
the objectives of both approaches, 'an understanding of the mature concepts
and theories of science and an understanding of the processes by which
scientific knowledge grows',[4] can be achieved simultaneously within a single,
coherent science curriculum. This last point will also involve a consideration
of the relationship between the knowledge taught in the science curriculum
and the knowledge that students acquire outside the schools in their everyday
life, knowledge which I shall refer to as commonsense knowledge.

2. SCIENCE AS PRODUCT VS. SCIENCE AS PROCESS

In favour of the presentation of science as a body of knowledge, it has been
argued that this knowledge represents a generally coherent picture of the
world and its phenomena which is as much a part of a student's cultural
heritage as the plays of Shakespeare or the novels of Balzac, and which can
be studied, like plays or novels, as a finished product. From this point of view

R. W. Home (ed.), Science Under Scrutiny, 159–172.
© 1983 by D. Reidel Publishing Company.

the informational content of science is emphasized as part of a student's general education and preparation for life in an industrialized society. The particular value of science, in addition to its general cultural function, is that it gives tremendous power over nature despite the abstract and highly mathematized form which much of its theory takes. The remoteness of scientific conceptual systems from everyday life is counterbalanced by the useful applications which these systems have in solving problems arising from everyday life. Finally, 'science is so vast that the students cannot acquire the most useful theories and problem-solving skills without undergoing a relatively dogmatic initiation'.[5] Consequently, a 'dogmatic element in science teaching, a logical presentation of objective knowledge is not poor practice by uninspired teachers but demanded by the structure of science'.[6] In summary, then, we may say that the presentation of science as a finished product is dogmatic and abstract, but useful for its general informational value and for its specific problem-solving potential.

In favour of the presentation of science as a process of discovery it has been argued that this approach not only teaches students 'to think about scientific things in the way that practicing scientists do',[7] but also cultivates in them an attitude of curiosity and an awareness of the fallibility of human knowledge. By its very nature a discovery approach to science grounds it in common experience and in commonsense ways of thought, since it begins with the students' 'spontaneous' reactions to certain phenomena and moves by a process of testing and refinement to a scientific conception of the same phenomena. Thus the discovery approach should lead students to an understanding of science as both a process and a product while avoiding dogmatism and undue abstraction.

The difficulty with the discovery approach in practice, however, is that 'only on the basis of the most superficial analysis of the nature of scientific activity, could it be said that such methods enable children to think and work in ways characteristic of a successful practitioner of science'.[8] In the first place the working scientist has an elaborated system of conceptual abstractions to make use of in formulating an interpretation of phenomena, whereas children must begin with their commonsense conceptions as a frame of reference.[9] Secondly, in the development of new scientific knowledge there is always a degree of uncertainty concerning the fit between theory and experience which must be resolved experimentally by the production of new phenomena, whereas in the discovery curriculum the experiments and phenomena are carefully contrived so as to lead to a predetermined outcome. Thus the contrast between the dogmatism of the product presentation and the 'openness' of the process presentation is not as great as is often claimed. As G. A. Ramsey expresses it, the contrast reduces to the following:

The learner may either be 'told' of Newton's useful models for understanding linear motion, using perhaps film or visual aid. Or on the other hand, the learner may be guided to inquire into motion, using the processes of science to learn the organizing structures Newton discovered and which have proved so useful in helping understand motion.[10]

In both cases it is Newton's laws that are to be learned and in both cases the criteria by which it has been decided that these laws are the most useful for understanding motion are not presented as needing justification. Finally, any curriculum based upon the discovery method must embody, either implicitly or explicitly, a philosophical theory about the nature of scientific methods and procedures. And since the methods for evaluating scientific evidence and philosophical arguments are entirely different,[11] it follows that science students will be even less able to evaluate critically the philosophical presuppositions of a discovery curriculum than to evaluate critically a dogmatically presented scientific theory.

So it would appear that the discovery approach is not significantly less dogmatic than the approach which presents scientific knowledge as a finished product, not does it give students experience with the actual process of scientific discovery and the production of new knowledge.

3. CAN THE OBJECTIVES OF BOTH APPROACHES BE ACHIEVED SIMULTANEOUSLY?

At this point it will be helpful to summarize briefly the main objectives of both of the approaches to science teaching which we have been considering, the product approach and the process approach. The main objective of the product approach is to initiate students as efficiently as possible into a tradition of problem-solving based upon a stable system of scientific concepts and practices, a tradition which T. S. Kuhn calls 'normal science'.[12] On the other hand the main objective of the process approach is to allow students to develop — out of their own experiences and in a way equivalent to the one pursued by a working scientist — a system of concepts and practices which converges on the system accepted within normal science. Because of this convergence, the system developed should share the problem-solving potential of normal science; but because of the students' understanding of how scientific knowledge grows, this system should be used critically rather than dogmatically and thus be open to major or minor revision at any time.

The incompatibility of the main objectives of the two approaches to science teaching has posed a problem for curriculum design since the inception of formal science education in the nineteenth century. While the effectiveness of the product approach in meeting its main objective has been demonstrated in practice, its unsuitability for meeting the main objective of the process approach is clear, since its function 'is not to produce sceptics, but to train highly competent "puzzle-solvers" who will work within the agreed framework of rules and theories'.[13] On the other hand, the effectiveness of the process approach in meeting even its own main objective is open to serious doubt, as we have argued above. For, in Layton's words, 'even with heuristic methods of learning it is difficult to avoid acquiring a misleading impression of science as a human activity. ... At the school level ... the acquisition of scientific knowledge is inescapably tinged with dogmatism'.[14] The question we must ask, then, is whether the main objectives of the two approaches to science education are necessarily incompatible, or whether a science curriculum is possible which at the same time leads to the efficient and systematic acquisition of scientific concepts, practices and problem-solving skills by students, and also reveals the process of production and development of these concepts and practices in such a way that students use them critically rather than dogmatically. I shall argue that such a curriculum is possible, but only if it relates scientific knowledge to commonsense knowledge in a way that is different from the relationship set up in both the product and the process approaches to science teaching.

4. COMMONSENSE KNOWLEDGE VS. SCIENTIFIC KNOWLEDGE

Commonsense knowledge arises directly from our daily experience of interacting both with everyday objects[15] and with other members of our society.[16] Our practical activities, involving physical manipulation and social exchanges, produce and reinforce in us beliefs and expectations about the nature of reality which may in some cases be given systematic expression but which are usually held tacitly and only articulated if they are challenged. Because of its immediate origin in everyday experience, commonsense knowledge is directly applicable to that experience, and its effectiveness as a guide to our practical activity in daily life is a measure of its success.[17] By contrast, however, scientific knowledge is highly abstract and arises from a specialized process of testing and refinement, which makes its conceptual content remote from our daily experience. When this scientific knowledge is related to the phenomena of everyday life, through a fairly elaborate process of applied research and engineering procedures, it can give us a powerful means of controlling these phenomena; nevertheless, many aspects of scientific knowledge are never suitable for application in this way. So we may say that commonsense knowledge is immediate and concrete, while scientific knowledge is remote and abstract.

There is, however, another contrast between these two forms of knowledge which I would like to draw here. Because of its immediacy and concreteness, commonsense knowledge is relatively static and changes only slowly, in response to social development. It has no history of its own, no inner dynamic of development, but tends to follow in a laggardly fashion behind changes in the social and technological context of everyday experience. This relative stability of commonsense knowledge, together with its immediacy and concreteness, builds an ontological commitment known as 'naive realism'[18] — in other words, we find it difficult to doubt whether tables, chairs, beer bottles, etc., really exist in the way that they seem to do. We take our commonsense knowledge not as a theoretical construct but as an accurate picture or description of stable reality. Abstract scientific knowledge, on the other hand, tends to break down ontological commitments: first of all because the specialized practices of science make this knowledge a tool for the production of new knowledge under controlled conditions; and secondly because the abstract objects of scientific thought are 'constituted' by the specification of operationally-defined properties and by the logical relations which they have to other abstract objects within a given theoretical structure. Thus within the context of scientific practice, the contents of scientific knowledge are neither stable nor a description of real objects.

It is true that in addition to the logical structure of a theory, scientific practice often makes use of a heuristic model or physical analogy which interprets or depicts abstract theoretical objects in a certain way; however the model is in no way an essential part of a scientific theory, as evidenced by the fact that some theories can be interpreted in a consistent way by different models which are incompatible with each other,[19] while other theories cannot be interpreted in a consistent way by any single model.[20] So although individual scientists may happen to regard a particular model as a description of real objects, this stance results from a psychological feeling of conviction on their part and not from any inherent theoretical necessity.[21] The heuristic model functions as a kind of 'mental crutch' which helps the imagination of the researcher to devise new experiments or to extend the theory in a new way; but it does not function as an accurate depiction of stable reality.

5. THE WORST OF BOTH WORLDS: SCHOOL SCIENCE

Now that I have characterized some of the differences between commonsense knowledge and scientific knowledge in this way, I shall attempt to show that school science, whether presented through a product approach or through a process approach, generally combines the worst aspects of both forms of knowledge, and that this is the reason why it encourages both dogmatism and a false conception of the production of scientific knowledge.

School science presents scientific knowledge as an abstract system of concepts and procedures divorced not only from daily life but also from the actual scientific practices by which that knowledge is produced. It thus represents scientific knowledge as relatively static, like commonsense knowledge, but without the immediacy and concreteness of the latter. School science has no history of its own and changes only in response to social and scientific developments. Whether school scientific knowledge is presented as a finished product or as a predetermined 'best' solution to a contrived problem, it has the stability of commonsense knowledge and is thus productive of ontological commitment in the form of belief in the reality of the abstract objects of scientific theory.

Moreover, prevailing pedagogical practice tends to reduce the distance between scientific abstraction and commonsense knowledge in a way that supports this ontological commitment by providing even simpler models of the already limited models used in scientific practice.[22] The abstractions of scientific knowledge are 'tied down' to elements of everyday experience through familiar examples and analogies, and thus take on the ontological status of commonsense knowledge; but they nevertheless remain abstractions from everyday life since the scientific concepts are not derived in a consistent way from everyday practice but instead the everyday experiences are selected to illustrate scientific concepts derived from a different sphere of practice altogether. Thus the collection of everyday experiences used to illustrate scientific knowledge has no coherence in itself and serves only to give students a feeling of familiarity with regard to scientific concepts which is completely illusory.

For example, the model or image of an electric current flowing through a garden hose serves to 'tie down' the abstract concept of current electricity by assimilating it into the familiar world of everyday experience. But this same image which creates in students a feeling of familiarity with the concept of current electricity also leads to the belief that if the switch at the power point is turned on when no appliance is plugged in, then electricity will leak out of the power point on to the floor. It must then be explained that electricity is not like water in this respect, that a circuit must be completed, that air is an insulator, and so on. A series of new images or models may be used to illustrate the concepts of circuit, conduction, and insulation, but these will be incompatible with the flowing water analogy used before. As more and more aspects of electrical theory are brought into consideration, more and more pedagogical models from everyday life may be used to explain them. But these everyday models will not be unified by one coherent form of everyday practice, such as the use of plumbing; they will only have the electrical theory itself as their principle of unity. It is for this reason that pedagogical models, although they give a tinge of everyday familiarity to scientific concepts, nevertheless remain abstractions from everyday life — abstractions in the sense that they are divorced from their normal context in everyday life. So instead of a coherent abstract theory with rigorously controlled application to certain specified phenomena, students acquire the notion of a literally chimerical substance, electricity, which exists for them in the same way that tables, chairs, beer bottles and garden hoses exist, which 'has' a collection of properties that are each familiar but are here combined in a hybrid way, and which 'behaves' in accordance with the 'nature' of these properties.

Now I submit that it is this ontological commitment, this belief in the reality of the abstract objects of scientific theory, which accounts for the dogmatic element in science teaching. When scientific abstractions take on the reality of everyday objects the possibilities for using them critically are drastically reduced. Furthermore, I also submit that this ontological commitment is in no way necessary to the students' efficient and systematic acquisition of scientific concepts, practices and problem-solving skills. In the same way that one can systematically learn a foreign language and use it effectively to describe phenomena without believing that it gives a truer description than one's native language, so too it should be possible to learn science as a kind of 'second language' for describing phenomena without acquiring ontological commitments in the process.[23] How, then, might this be done?

6. FROM CONFLATION TO CONFRONTATION

I suggest that instead of attempting to minimize the difference between scientific abstractions and commonsense knowledge, we should present these two forms of knowledge in direct confrontation with each other. As

Paul Feyerabend has emphasized, it is only when alternative systems of knowledge confront one another that the hidden assumptions of both systems are
made explicit and therefore become subject to criticism and defence.[24] This
confrontation requires that neither system be prejudged as 'correct' in
advance; hence the point of the confrontation must not be to show that one
form of knowledge is 'truer' than another, but rather to examine both kinds
of knowledge to determine the ways in which they are produced and their
legitimate domains of application.

Feyerabend has recommended a confrontation of science with non-
scientific systems such as voodoo, witchcraft and astrology, all of which he
would have taught in the schools on an equal basis.[25] This suggestion, like my
own, would lead to the breakdown of ontological commitments in science
teaching. There are however, at least three reasons why I consider my own
proposal, of a confrontation between science and commonsense, to be
superior. The first reason is pragmatic: leaving aside the question of the
desirability of teaching voodoo, witchcraft and astrology in the schools, it is
completely 'utopian' to expect that such knowledge would become part of
the formal curriculum in the schools of our industrial, high-technology
society, because this knowledge arises from practices which have no 'recognized' place in this society. Commonsense knowledge, however, arises
from the everyday practices of our society and could therefore more easily
become integrated with the curriculum in a formal way.[26] The second reason
in favour of my proposal is pedagogical: the knowledge systems of voodoo,
witchcraft and astrology are as theoretically complex and conceptually
remote from the experience of most people in our society as is the knowledge
system of science, and therefore as difficult to master; whereas commonsense
knowledge on the other hand is immediate and concrete, and has already
been acquired by most people in the course of their daily experience. Finally,
my third reason is critical: the confrontation I am suggesting would tend to
reveal our commonsense knowledge as a system of constructs dependent on
our form of society, with no special claim to be the 'true' depiction of reality.
By thus depriving commonsense knowledge of its ontological commitment we
would open the way for a more critical analysis of our form of society, since
the ontology of commonsense knowledge often justifies existing social relations by making them seem natural or inevitable. On the other hand, the
process of depriving voodoo, witchcraft and astrology of their ontological
commitments would do very little to facilitate students' critical thinking
about their own society.

For an example of how my proposal might be put into practice, we can return to the teaching of electricity. Instead of attempting to reduce the scientific concepts of electrical theory to commonsense knowledge by drawing upon analogies from areas of everyday life unrelated to the experience of electricity, such as the use of garden hoses, we might start with everyday experiences involving electrical phenomena themselves. A good starting point might be the experience of turning a switch in one part of a room and having a light go on in some other part of the room. Students would be asked to explain this experience in terms of their commonsense ideas about the phenomena involved. To this extent our point of departure resembles that of the discovery curriculum; however, unlike the discovery approach, this approach would not seek to 'derive' scientific concepts about electricity from the students' commonsense ideas — it would not treat these ideas as first approximations to a scientific theory. Instead, it would present the relevant aspects of physical theory as a wholly different form of explanation of the phenomena in question. Thus presented, as an element of a theoretical system rather than as a series of images, analogies or models, the scientific explanation of these phenomena would be unlikely to have much intuitive appeal; but this feature should be regarded as a strength rather than as a weakness, for it immediately raises the question of how such a theoretical system could come to be developed and accepted. In other words, a direct presentation of unfamiliar scientific concepts (at the level of formalization appropriate to the students' degree of maturity) opens the way to a discussion of the scientific practice which produces and employs these concepts, and of the differences between this practice and the common practices of everyday life. By contrast, the presentation of the scientific concepts through the medium of familiar, everyday images such as garden hoses, tends to close off this line of inquiry because of the specious 'obviousness' which it imparts to these concepts.

The approach to science teaching which I am advocating here — namely, the confrontation of commonsense knowledge and scientific knowledge in order to determine the forms of production and the legitimate domains of application of each — requires that we develop an instrumentalist presentation of knowledge rather than a naive realist presentation. In fact, it requires that we develop a pedagogy based on the view which Alan Chalmers, in his book *What Is This Thing Called Science?*, has described as 'radical instrumentalism'.[27] This view is radical because, unlike classical instrumentalism, it applies to all knowledge — scientific and commonsense alike — and it is instrumentalist because it treats this knowledge not as a description of reality but as an intellectual tool or instrument used for specific purposes in specific forms of practice and produced for those purposes by means of specific forms of practice.[28]

Thus, for example, a scientific theory can be seen as a tool used within the specialized activities of scientific practice for the purpose of interpreting, manipulating and producing specific experimental phenomena. The same theory, in combination with other knowledge, can also be used as a tool within engineering practice for the purposes of technological development. Similarly, commonsense ideas can be seen as tools used within the context of everyday activities for the purposes of successfully interacting with common objects and with other members of our society, so as to accomplish the aims of daily life. In all these cases the evaluation of knowledge must start not from the question, 'Does it give a true picture of reality?', but from questions such as 'Within what practice(s), by what means, and for what purpose(s) is this knowledge used as a tool?', and 'From what practice, by what means, and for what purpose was this knowledge developed?'. Finally, since purposes are always dependent upon individual or social interests, we must include the question, 'In furtherance of what interests has this knowledge been produced and employed?'

A pedagogical approach which started from such questions as these, and applied them both to the explicit theories of scientific knowledge and to the largely implicit theories of commonsense knowledge, would lead to an investigation of (a) the different content of these two forms of knowledge, (b) the different practices out of which they arise, (c) the different interests and purposes to which they relate and the different forms of human activity in which they find application, and finally (d) the different degrees to which particular theories employed within a given practice can fulfill the purposes for which they are employed.

An approach of the sort suggested here would not attempt to separate the study of the methodology and social processes of science from the study of the content of science, as has sometimes been suggested to resolve the process/product dilemma in science education.[29] It would, on the contrary, provide a perspective for the unified study of knowledge, in both its scientific and its commonsense forms, and of social relations, in both the specialized area of scientific activity and the general area of everyday life. It would, in other words, provide the basis for an 'integrated curriculum', in the sense defined by Basil Bernstein,[30] which would involve the blurring of traditional disciplinary boundaries and probably also the introduction of team teaching. The practical difficulties facing the successful operation of such an integrated curriculum, combining social studies and natural science, would be formidable; but the experience gained elsewhere in the use of other integrated curricula can serve as at least a partial guide in this respect.

Apart from the practical difficulties, however, it may seem that the approach outlined here, while accomplishing the main objective of the process approach to science education, would not be able to accomplish the main objective of the product approach: namely, the training of skilled problem-solvers within the tradition of existing normal science. But I would answer that the analysis of the practices by which scientific knowledge is produced, the purposes for which it is used, and the degree to which it successfully fulfills · those purposes, would necessarily involve the use by students of scientific knowledge in problem-solving exercises. In order to understand the practices and purposes of normal science, students would have to engage in the activities appropriate to normal science. In order to ascertain the limits of the applicability of commonsense knowledge and of scientific knowledge, they would have to apply these forms of knowledge in the appropriate types of practice. But with the students being conscious of these limits and with both commonsense knowledge and school scientific knowledge being deprived of their ontological commitment, the theoretical content of normal science would have no dogmatic force but would function simply as a kind of tool-box from which tools can be drawn for use in certain contexts.[31] In this process neither scientific knowledge nor commonsense knowledge would be devalued, but each would be assigned its appropriate place in the student's intellectual repertoire.

7. SOME SOCIAL IMPLICATIONS

Having shown, I hope, the possibility of resolving the process/product dilemma in science teaching, I would like to conclude by pointing to some other results which would follow from the type of curriculum I have suggested and which I consider socially desirable. First, by bringing science and commonsense into confrontation rather than attempting to conflate the two, commonsense ideas would be less likely to serve as 'epistemological obstacles' or intellectual barriers to the development of new scientific knowledge.[32] But conversely, and more importantly, scientific ideas could less easily serve to justify elements of commonsense knowledge based on unjust social relations: for example, ideas drawn from rigorous laboratory experiments in genetics could less easily be used to justify commonsense ideas about sex roles or racial characteristics of humans, if it were well understood that these ideas derive from two distinct forms of practice and are applicable only within the specific limits of those practices. Furthermore, understanding how the commonsense ideas in question derive from specific existing social relations, and analyzing the purposes and interests served by these ideas and relations, could lead students to consider what changes in these social relations would serve their own purposes and interests, and what forms of practice and knowledge would help to bring these changes about.

This process of opening students' minds to their own potentialities, and providing them with the intellectual tool-boxes from which they can choose their own instruments as needed, I consider to be the proper function of teaching. I distinguish teaching here from 'preaching', which I would define broadly as the process of producing convictions about reality in the minds of students, whether by imparting to them the ontological commitments of school scientific knowledge or by reinforcing the ontological commitments of their uncritical commonsense knowledge.[33] On this definition, much of what now occurs in science education would have to be classified as science preaching rather than science teaching, not because of the intentions of teachers but because of the structure of science education itself. I hope that I have indicated a way in which this structure can be changed to eliminate science preaching while allowing science teaching to flourish.

University of New South Wales

NOTES

[1] Layton, David (1973) *Science for the People: The Origins of the School Science Curriculum in England*, London: Allen and Unwin, p. 174.
[2] Ramsey, G. A. (1975) 'Science Teaching as an Instructional System', in P. L. Gardner (ed.), *The Structure of Science Education*, Hawthorn: Longman Australia, p. 96.
[3] *Ibid., op. cit.*, p. 97.
[4] Layton, *op. cit.*, p. 176.
[5] Butler, J. E. (1977) 'Radical Philosophy and History of Science', *Australian Science Teachers Journal* 23(2), 40. It should be noted that Butler is critical of the view represented by this quotation and the one which immediately follows.
[6] *Ibid.*
[7] Layton, *op. cit.*, p. 175.
[8] *Ibid.*
[9] Cf. the criticism of one 'extreme example' of a process curriculum by Myron Atkin, quoted in Layton, *op. cit.*, p. 174. While the curriculum in question may be an extreme example, the problem of the students' conceptual starting point must be faced in designing any process curriculum.
[10] Ramsey, *op. cit.*, p. 97.

[11] Lucas, Arthur M. (1977) 'Should "Science" be Studied in Science Courses?' *Australian Science Teachers Journal* **23**(2), 31–37.

[12] Kuhn, T. S. (1970) *The Structure of Scientific Revolutions*, 2nd ed., Chicago: University of Chicago Press. Cf. Biggins, David R., and Ian Henderson (1978) 'What is Science Teaching For?' *Physics Education* **13**, 438–441; and Butler, *loc. cit.*

[13] Butler, *loc. cit.*

[14] Layton, *op. cit.,* p. 176.

[15] Cf. Chalmers, Alan F. (1976) *What Is This Thing Called Science?* St. Lucia: University of Queensland Press, pp. 130–131.

[16] Cf. Berger, Peter, and Thomas Luckmann (1966) *The Social Construction of Reality*, Harmondsworth: Penguin.

[17] Since different forms of society (and within a given society, different classes) have different criteria for what counts as 'effective practical activity', it is clear that the content of commonsense knowledge will be specific to each form of society (and class).

[18] Chalmers, *loc. cit.;* cf. pp. 113–115.

[19] For example, Fourier's analytical theory of heat is consistent with either a kinetic model or a fluid model of heat; cf. Friedman, Robert M. (1977) 'The Creation of a New Science: Joseph Fourier's Analytical Theory of Heat', *Historical Studies in the Physical Sciences* **8**, 73–99.

[20] For example, the quantum theory of light cannot be interpreted consistently by either a wave model or a particle model.

[21] Cf. Roqueplo, Philippe (1974) *Le partage du savoir*, Paris: Seuil, pp. 100–103.

[22] Roqueplo refers to this process as the 'overmodelling' (*surmodélisation*) of the scientific model (*op. cit.,* pp. 109, 141). While Roqueplo's principal concern is the popularization of science in the mass media, nevertheless many of his observations are equally applicable to the context of school science teaching.

[23] The presentation of science to non-Western children 'as a "second culture," valid in its own right and taught in much the same spirit as a second language is taught', was suggested by Francis E. Dart (1972) 'Science and the Worldview', *Physics Today* **25**(6), 54 (cf. *idem.* (1973) 'The Cultural Context of Science Teaching', *Search* **4**, 326). The present proposal extends this suggestion to the teaching of Western children, whose 'native culture' is the everyday experience of industrial society.

[24] Feyerabend, Paul K. (1975) *Against Method*, London: New Left Books.

[25] *Ibid.*

[26] I say 'in a formal way' because commonsense knowledge is already incorporated throughout the school curriculum in an informal way, as an unanalyzed component. The proposal advanced here is that commonsense knowledge itself should formally become an object of study within the curriculum, to be analyzed critically together with scientific knowledge.

[27] Chalmers, *op. cit.,* pp. 129–134.

[28] Chalmers also describes his view as 'pluralistic realism', for the following reason: 'The radical instrumentalist or pluralistic realist wishes to emphasize the distinction between our conceptual systems, whether they be scientific theories or those presupposed in everyday language and which are human products that are subject to change, and the real world to which those real conceptual systems bear some relation. Both scientific theories and the external world are real, but they are not to be identified. ... Scientific theories are constantly produced and modified as a result of scientific

practice. The reasons behind my wishing to call this version of realism "pluralistic" should now be evident. The external world and the world of theories are both real, but they are distinct. They are linked by a third real, scientific practice.' (*Op. cit.*, pp. 129–130.)

[29] e.g., Lucas, *op. cit.*

[30] Bernstein, Basil (1971) 'On the Classification and Framing of Educational Knowledge', in M. F. D. Young (ed.), *Knowledge and Control*, London: Collier Macmillan, pp. 47–69; also reprinted in E. Hopper (ed.) (1971) *Readings in the Theory of Educational Systems*, London: Hutchinson, pp. 184–211.

[31] Cf. the comments of Michel Foucault on 'Theory as a Tool-Box' in Meaghan Morris and Paul Patton (eds.) (1979) *Michel Foucault: Power, Truth, Strategy,* Sydney: Feral Publications, p. 57.

[32] The theory of epistemological obstacles is presented, together with historical illustrations, in Bachelard, Gaston (1938) *La formation de l'esprit scientifique*, Paris: Vrin.

[33] There are also other ways, not directly dependent upon ontological commitments, by which science education produces convictions about reality in the minds of students. I have outlined a number of these processes in 'Some Social Effects of Science Teaching', in David R. Oldroyd (ed.) (1978) *Historical, Philosophical and Social Perspectives of Science in Secondary Education,* Kensington: University of New South Wales Press, pp. 36–40.

NOTES ON CONTRIBUTORS

RANDALL ALBURY is Associate Professor in the School of History and Philosophy of Science at the University of New South Wales. A native of the United States, he received his Ph.D. in the history of science from the Johns Hopkins University in Baltimore, Maryland, before moving to Sydney in 1973. He has published in most of the leading history of science journals, chiefly but by no means exclusively on topics in the history of biology. His annotated translation of Etienne de Condillac's *Logic* (1780) was published in 1980. He is also the author of *The Politics of Scientific Objectivity*, published by Deakin University Press in 1983.

LLOYD EVANS, who opened the symposium that gave rise to the papers comprising this volume as the then (1978–1982) president of the Australian Academy of Science, was born in New Zealand and took his first degrees there before going to Oxford as a Rhodes scholar. After completing his doctorate in Oxford, he worked in the phytotron at the California Institute of Technology, as a Harkness Fellow, before joining the CSIRO Division of Plant Industry in Canberra in 1956. He is a plant physiologist with a strong involvement in international agricultural research, and with an active interest in the history of science. Chief of his Division, 1971–1978, he was elected a Fellow of the Australian Academy of Science in 1971 and of the Royal Society of London in 1976.

BRYAN GANDEVIA has since 1963 been associate professor of medicine and chairman of the Department of Respiratory Medicine at the Prince Henry and Prince of Wales Hospitals, Sydney, and the University of New South Wales. His interest in the history of medicine goes back to his student days at the University of Melbourne. In addition to numerous scientific publications in his field of medical specialization, he has published *An Annotated Bibliography of the History of Medicine in Australia* (1957), *Occupation and Disease in Australia Since 1788* (1971), *Tears Often Shed: Child Health and Welfare in Australia Since 1788* (1978), and medico-historical articles in a variety of Australian and international medical and historical journals.

R. W. Home (ed.), Science Under Scrutiny, 173–175.

ROM HARRÉ was born in New Zealand and educated at the University of Auckland (mathematics and engineering) and the University of Oxford (philosophy). He has been University Lecturer in Philosophy of Science and Fellow of Linacre College, Oxford, since 1960, and Adjunct Professor of Social and Behavioral Science, State University of New York at Binghamton, since 1973. He is co-founder of the *Journal for the Theory of Social Behaviour* and author of numerous books including *The Principles of Scientific Thinking* (1970), *The Explanation of Social Behaviour* (with P. F. Secord) (1972), *Causal Powers* (with E. H. Madden) (1975) and *Social Being: A Theory for Social Psychology* (1979).

EVERETT MENDELSOHN teaches the history of science at Harvard University, where he pursues the history of the biological sciences and the social/sociological history of science. He edits the *Journal of the History of Biology* and is an active member of the Editorial Board of the Yearbook, *Sociology of the Sciences*. He has had a longstanding interest in contemporary relations of science and society at both the scholarly level (Vice-President of the International Council for Science Policy Studies) and the activist level, helping to found the A.A.A.S. Committee on Science, Arms Control, and National Security. A deep interest in the Middle East and the cause of reconciliation lay behind his most recent book (a report prepared for the American Friends Service Committee — Quakers), *A Compassionate Peace, A Future of the Middle East*, New York: Hill and Wang; London: Penguin, 1982.

ALAN MUSGRAVE was born in England in 1940, studied and taught at the London School of Economics from 1958 to 1970, and is currently Professor of Philosophy at the University of Otago. Together with Imre Lakatos he edited *Problems in the Philosophy of Science* (1968) and *Criticism and the Growth of Knowledge* (1970), and has published numerous papers in Philosophy and History of Science.

JOHN PASSMORE, born and educated in Sydney, is Emeritus Professor of Philosophy and Visiting Fellow in History of Ideas at the Australian National University, Canberra. He is a Fellow and past President of the Australian Academy of the Humanities, Fellow of the Academy of the Social Sciences in Australia and a former Vice-President of the Institut International de Philosophie. His books include *Hume's Intentions* (1952), *A Hundred Years of Philosophy* (1957), *Man's Responsibility for Nature* (1974), *Science and Its Critics* (1978) and *The Philosophy of Teaching* (1979).

JARLATH RONAYNE has been Professor of History and Philosophy of Science at the University of New South Wales since 1977. A graduate in natural sciences of Trinity College, Dublin, he received his doctoral training in chemistry at the University of Cambridge before turning to problems in the science policy area. He is author of papers in chemical physics and of *Science in Government: An Introduction to the Principles and Practice of Science Policy* (1983), and co-edited *Science, Technology and Public Policy* (1979). He has been consultant and visiting advisor to the Secretariat of the Australian Science and Technology Council (1978—81) and is currently a consultant to the Australian government's Department of Science and Technology, engaged on a study of the feasibility of introducing a science indicators program in Australia.

HUGH STRETTON is Reader in History at the University of Adelaide and a Fellow of both the Australian Academy of the Humanities and the Academy of the Social Sciences in Australia. His publications include *The Political Sciences* (1969), *Ideas for Australian Cities* (1970), *Housing and Government: The 1974 Boyer Lectures* (1974), *Capitalism, Socialism and the Environment* (1976) and *Urban Planning in Rich and Poor Countries* (1978)

The Editor
ROD HOME has been Professor of History and Philosophy of Science at the University of Melbourne since 1975. He graduated in science from the University of Melbourne before completing a Ph.D. in the history of science at Indiana University in 1967. His publications have been chiefly concerned with aspects of the history of physics, and include *Aepinus's Essay on the Theory of Electricity and Magnetism* (with P. J. Connor) (1979) and *The Effluvial Theory of Electricity* (1981).

INDEX

DATE DUE

DEMCO 38-297